Problem Solvers

Edited by L. Marder
Senior Lecturer in Mathematics, University of Southampton

No. 13

Electromagnetism

Problem Solvers

1. ORDINARY DIFFERENTIAL EQUATIONS — J. Heading
2. CALCULUS OF SEVERAL VARIABLES — L. Marder
3. VECTOR ALGEBRA — L. Marder
4. ANALYTICAL MECHANICS — D. F. Lawden
5. CALCULUS OF ONE VARIABLE — K. Hirst
6. COMPLEX NUMBERS — J. Williams
7. VECTOR FIELDS — L. Marder
8. MATRICES AND VECTOR SPACES — F. Brickell
9. CALCULUS OF VARIATIONS — J. W. Craggs
10. LAPLACE TRANSFORMS — J. Williams
11. STATISTICS I — A. K. Shahani and P. K. Nandi
12. FOURIER SERIES AND BOUNDARY VALUE PROBLEMS — W. E. Williams
13. ELECTROMAGNETISM — D. F. Lawden

Electromagnetism

D. F. LAWDEN

Professor of Mathematical Physics
University of Aston in Birmingham

LONDON · GEORGE ALLEN & UNWIN LTD

RUSKIN HOUSE MUSEUM STREET

First published in 1973

ISBN 0 04 538001 5 *hardback*
0 04 538002 3 *paperback*

Printed in Great Britain
in 10 on 12 pt 'Monophoto' Times Mathematics Series 569
by Page Bros (Norwich) Ltd., Norwich

Contents

Preface

This text covers the whole range of the classical theory of electromagnetism from Coulomb's Law to Maxwell's equations and is therefore suitable for use by students of applied mathematics, physics or electrical engineering at all stages of their university courses. It is assumed that such students will be familiar with the notation and important theorems of vector calculus as, for example, contained in two other texts from this series, *Vector Algebra* and *Vector Fields* by L. Marder. It is understood that all physical quantities are expressed in SI units, and, for convenience, a list of these is provided in an Appendix.

My secretary, Mrs A. Breakspear, prepared the typescript, a difficult task for which I am most grateful.

D. F. LAWDEN

University of Aston,
February, 1972

Chapter 1

The Electrostatic Field

1.1 Field Intensity An electrostatic field is described by specifying the electric intensity vector $\boldsymbol{E}$ at every point. $\boldsymbol{E}$ is the force per unit charge. Thus, if a point charge q is placed at a point where the electric intensity is $\boldsymbol{E}$, it experiences a force $\boldsymbol{F}$ given by

$$\boldsymbol{F} = q\boldsymbol{E}. \tag{1.1}$$

The intensity at P due to a single point charge q at O acts along OP and has magnitude which is inversely proportional to OP^2. Thus, if $\text{OP} = \boldsymbol{r}$ and $\hat{\boldsymbol{r}}$ is the unit vector along OP, then

$$\boldsymbol{E} = \frac{1}{4\pi\varepsilon_0}\cdot\frac{q}{r^2}\hat{\boldsymbol{r}} = \frac{1}{4\pi\varepsilon_0}\cdot\frac{q}{r^3}\boldsymbol{r}. \tag{1.2}$$

In this equation, all quantities are measured in SI units (see p. 95) and $\varepsilon_0 = 8{\cdot}85 \times 10^{-12}$ is called the *permittivity of free space.*

It follows from equations (1.1) and (1.2) that the force exerted upon a charge q' at P by the charge q at O is given by

$$\boldsymbol{F} = \frac{1}{4\pi\varepsilon_0}\cdot\frac{qq'}{r^2}\hat{\boldsymbol{r}}. \tag{1.3}$$

This is *Coulomb's Law.*

The electric intensity due to any given distribution of charge can be calculated by vector addition of the intensities due to the individual charges.

***Problem* 1.1** The surface of a disc of ebonite is uniformly charged. Calculate the electric intensity at points on the disc's axis.

Solution. Let σ be the charge per unit area on the disc. Consider the contribution to the intensity at P (Fig. 1.1) of the elemental ring of radius r and thickness dr. Any charge q at a point Q on this element generates an intensity $q/4\pi\varepsilon_0\,\text{PQ}^2$ at the point P in the direction QP. It is clear from the symmetry that the resultant intensity at P is directed along OP; this implies that only the component along OP of $q/4\pi\varepsilon_0\,\text{PQ}^2$, namely $q\cos\theta/4\pi\varepsilon_0\,\text{PQ}^2$, contributes to this resultant. Thus, the net contribution of the elemental ring is $(\sum q)\cos\theta/4\pi\varepsilon_0\,\text{PQ}^2$, where $\sum q$ is the total charge on the ring. But $2\pi r\,dr$ is the area of the ring, so that $\sum q = 2\pi\sigma r\,dr$.

Integrating over all the rings, we calculate that the overall intensity at P has magnitude

$$E = \int_0^a \frac{\sigma r \cos\theta}{2\varepsilon_0\, \mathrm{PQ}^2} dr.$$

If OP $= z$, then $r = z\tan\theta$, $dr = z\sec^2\theta\, d\theta$ and PQ $= z\sec\theta$. Hence

$$E = \frac{\sigma}{2\varepsilon_0}\int_0^\alpha \sin\theta\, d\theta = \frac{\sigma}{2\varepsilon_0}(1-\cos\alpha), \tag{1.4}$$

where α is the angle subtended by a radius of the disc at P. □

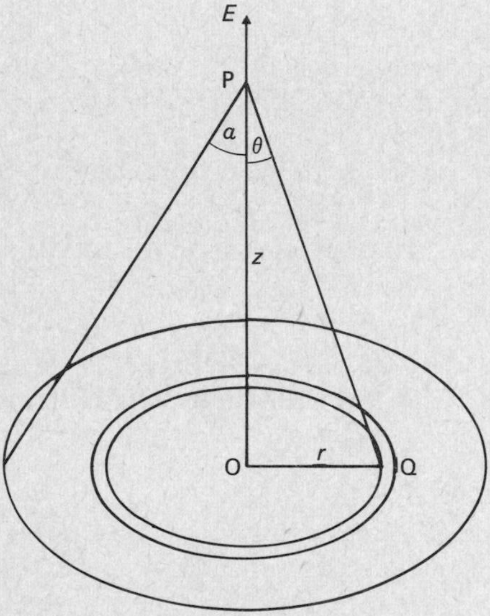

Fig. 1.1

***Problem* 1.2** A very long straight rod is uniformly charged. Calculate the electric intensity at a distance ρ from the rod.

Solution. The intensity at a point P (Fig. 1.2) must be directed along the perpendicular OP to the rod. Let Q be a point on the rod distant x from O and let QQ′ $= dx$ be an element of the rod. Then, if q is the charge per unit length, the intensity at P due to the element has magnitude $q\, dx/4\pi\varepsilon_0\, \mathrm{PQ}^2$ and is directed along QP. Only the component in the direction OP contributes to the resultant intensity at P. It follows by integration that this resultant is given by

$$E = \int_{-\infty}^{\infty} \frac{q\cos\theta}{4\pi\varepsilon_0\,\mathrm{PQ}^2}\,dx.$$

Since $x = \rho\tan\theta$, $dx = \rho\sec^2\theta\,d\theta$, $\mathrm{PQ} = \rho\sec\theta$, this integral is equivalent to

$$E = \frac{q}{4\pi\varepsilon_0\,\rho}\int_{-\frac{1}{2}\pi}^{\frac{1}{2}\pi}\cos\theta\,d\theta = \frac{q}{2\pi\varepsilon_0\,\rho}. \qquad \square \quad (1.5)$$

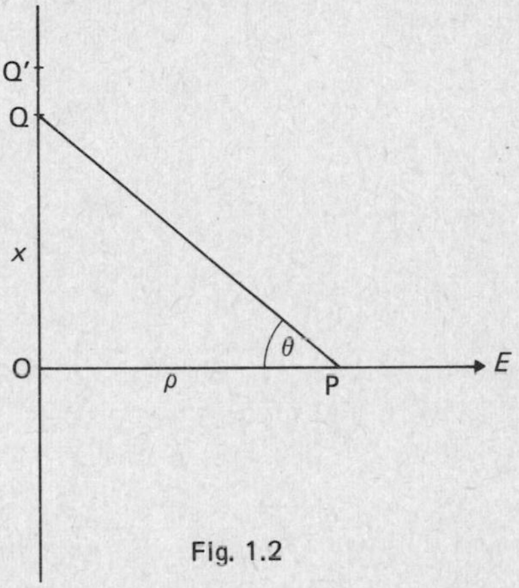

Fig. 1.2

***Problem* 1.3** An ebonite rod is uniformly charged, the total charge being q. It is placed along the axis of a disc of ebonite, also uniformly charged with charge Q, so that one end of the rod is at the centre of the disc. If the radius of the disc and the length of the rod are both a, show that the rod is repelled by a force $Qq(2-\sqrt{2})/(2\pi\varepsilon_0\,a^2)$.

Solution. Consider an element dx of the rod, distant x from the centre of the disc. The charge on the element is $q\,dx/a$; it lies in a field of intensity $Q(1-\cos\theta)/(2\pi\varepsilon_0\,a^2)$ generated by the disc (see equation (1.4)), where θ is the angle subtended by a radius of the disc at the element. Thus, the element is repelled along the disc's axis by a force $Qq(1-\cos\theta)\,dx/(2\pi\varepsilon_0\,a^3)$. Since $\cos\theta = x/\sqrt{(a^2+x^2)}$, by integration with respect to x from $x = 0$ to $x = a$, the resultant force on the rod is now easily found to be that stated. □

1.2 Potential When a point charge is taken from one point to another in an electrostatic field, the work done by the field force acting on the charge is always independent of the path joining the points. Any field for which this principle holds is said to be *conservative*.

The work done by the field force when a point charge is taken from a point P to a fixed datum point D is called the *potential energy* of the charge at P. The potential energy of unit charge at P is called the *potential* of the field at P and will be denoted by V.

It is usual to choose the datum point to be at a great distance from the charges responsible for the field. In the case of the field due to a single point charge q, the force exerted on unit charge distant r from q is $q/(4\pi\varepsilon_0 r^2)$; thus, the work done by this force when the unit charge recedes to an infinite distance is given by

$$V = \frac{q}{4\pi\varepsilon_0 r}. \tag{1.6}$$

This is the potential at distance r from q. It follows from equation (1.2) that, in this case,

$$\boldsymbol{E} = -\operatorname{grad} V. \tag{1.7}$$

Since every electrostatic field is generated by point charges (electrons, protons, etc.), this equation is valid for all such fields.

***Problem* 1.4** Calculate the potential of an infinite line charge q per unit length.

Solution. The electric intensity has been calculated in Problem 1.2. It is perpendicular to the line charge and its magnitude is given by equation (1.5). Clearly, V depends only on the distance ρ from the line charge. Also, by equation (1.7), the derivative of V in the direction of ρ must equal $-E = -q/2\pi\varepsilon_0\rho$. Hence integrating,

$$V = -\frac{q}{2\pi\varepsilon_0}\log\rho. \tag{1.8}$$

The constant of integration is taken to be zero, implying that the datum is at $\rho = 1$. Since $V \to -\infty$ as $\rho \to \infty$, the datum cannot be taken at infinity in this case. □

A *dipole* comprises point charges $+q$ and $-q$ at points A, A′ (Fig. 1.3). Their distance apart A′A $= d$ is assumed to be small and q is assumed to be correspondingly large so that the product qd is finite; this product is termed the *moment* of the dipole and will be denoted by m. The moment is usually taken to be a vector quantity, its direction being that of the dipole axis A′A; it will then be denoted by $\boldsymbol{m}$. By summing the contributions to the potential at P of the two charges and approximating to the first order in d, it is found that

$$V = \frac{m\cos\theta}{4\pi\varepsilon_0 r^2} = \frac{\boldsymbol{m}.\boldsymbol{r}}{4\pi\varepsilon_0 r^3} = -\frac{1}{4\pi\varepsilon_0}\boldsymbol{m}\cdot\text{grad}\left(\frac{1}{r}\right), \tag{1.9}$$

where (r, θ) are polar coordinates of P with pole O at the mid point of A′A. □

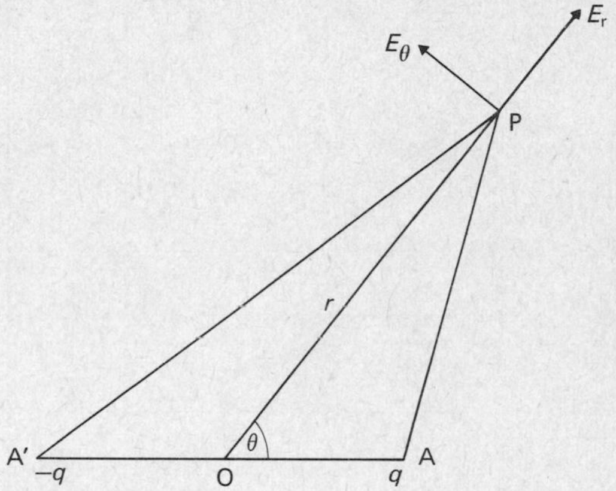

Fig. 1.3

***Problem* 1.5** Calculate the field intensity due to a dipole.

Solution. Let E_r, E_θ be the components of $\boldsymbol{E}$ in the radial and transverse directions at P (Fig. 1.3), i.e. along OP and perpendicular to OP. Then, by equation (1.7),

$$E_r = -\frac{\partial V}{\partial r}, \qquad E_\theta = -\frac{1}{r}\frac{\partial V}{\partial \theta}. \tag{1.10}$$

Substituting for V from equation (1.9). we find

$$E_r = \frac{1}{4\pi\varepsilon_0}\cdot\frac{2m\cos\theta}{r^3}, \qquad E_\theta = \frac{1}{4\pi\varepsilon_0}\cdot\frac{m\sin\theta}{r^3}. \tag{1.11}$$

This result can also be expressed in vector form thus: We note that the vector $\boldsymbol{m}/4\pi\varepsilon_0 r^3$ has components $(m\cos\theta/4\pi\varepsilon_0 r^3,\ -m\sin\theta/4\pi\varepsilon_0 r^3)$ in the radial and transverse directions. It follows from equations (1.11) that the vector $\boldsymbol{E}+\boldsymbol{m}/4\pi\varepsilon_0 r^3$ has zero transverse component and radial component $3m\cos\theta/4\pi\varepsilon_0 r^3$. Hence we can write

$$\boldsymbol{E}+\frac{1}{4\pi\varepsilon_0}\cdot\frac{\boldsymbol{m}}{r^3} = \frac{1}{4\pi\varepsilon_0}\cdot\frac{3m\cos\theta}{r^4}\boldsymbol{r}$$

$$= \frac{1}{4\pi\varepsilon_0}\cdot\frac{3\boldsymbol{m}\cdot\boldsymbol{r}}{r^5}\boldsymbol{r}.$$

Thus

$$\boldsymbol{E} = \frac{1}{4\pi\varepsilon_0}\left(\frac{3\boldsymbol{m}\cdot\boldsymbol{r}}{r^5}\boldsymbol{r}-\frac{\boldsymbol{m}}{r^3}\right). \qquad \square(1.12)$$

***Problem* 1.6** A particle of mass M and charge q is free to move in a plane containing the axis of the dipole shown in Fig. 1.3. At $t = 0$, it is projected from the point $r = a$, $\theta = \alpha$, with transverse velocity V and zero radial velocity. Prove that, at any later time t, $r^2 = a^2+kt^2$, where k is a constant. Hence find the condition for the particle to describe a circle.

Solution. The energy equation for the particle is

$$\tfrac{1}{2}M(\dot{r}^2+r^2\dot{\theta}^2)+\frac{qm\cos\theta}{4\pi\varepsilon_0 r^2} = \tfrac{1}{2}MV^2+\frac{qm\cos\alpha}{4\pi\varepsilon_0 a^2}.$$

Using the first of equations (1.11), the radial component of the particle's equation of motion can be written

$$M(\ddot{r}-r\dot{\theta}^2) = \frac{2mq\cos\theta}{4\pi\varepsilon_0 r^3}.$$

Eliminating $\dot{\theta}^2$ between these equations, we derive the equation

$$r\ddot{r}+\dot{r}^2 = 2k \qquad (1.13)$$

where

$$k = \tfrac{1}{2}V^2+\frac{qm\cos\alpha}{4\pi\varepsilon_0 Ma^2}.$$

The left-hand member of equation (1.13) is $d(r\dot{r})/dt$. Hence, integrating under the initial condition $\dot{r} = 0$ at $t = 0$, we find that $r\dot{r} = 2kt$. Since $r\dot{r} = d(\frac{1}{2}r^2)/dt$, a further integration under the initial condition $r = a$ at $t = 0$ yields the stated equation.

The condition for a circular path is clearly $k = 0$, i.e.

$$m = -2\pi\varepsilon_0 MV^2a^2/(q\cos\alpha). \qquad \square$$

***Problem* 1.7** Charge is uniformly distributed with density ρ over the interior of a hemisphere of radius b. Calculate the potential at a point on the axis of the hemisphere, on the side remote from the plane face, at distance y from the centre. If a particle of mass m and charge $-2\pi\rho b^3/3$ is released from rest at a point on the axis where y has a large value, find the velocity with which it arrives at the curved surface.

Solution. If a thin disc of radius a has a uniform charge σ per unit area, it follows from equation (1.4) that the electric intensity at a point on its axis distant $z(> 0)$ from the centre is given by

$$E = \frac{\sigma}{2\varepsilon_0}\left[1 - \frac{z}{(z^2+a^2)^{\frac{1}{2}}}\right].$$

If V is the potential, then equation (1.7) requires that $E = -dV/dz$. Hence

$$V = \frac{\sigma}{2\varepsilon_0}\left[(z^2+a^2)^{\frac{1}{2}} - z\right]. \tag{1.14}$$

(N.B. This makes $V \to 0$ as $z \to \infty$.)

Dividing the hemisphere into disc-shaped elements of thickness dx as indicated in Fig. 1.4, x being the distance of an element from the centre O,

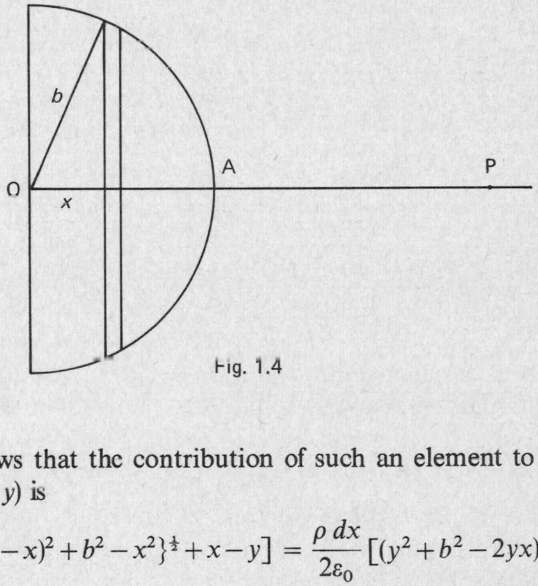

Fig. 1.4

it now follows that the contribution of such an element to the potential at P (OP $= y$) is

$$\frac{\rho\,dx}{2\varepsilon_0}[\{(y-x)^2+b^2-x^2\}^{\frac{1}{2}}+x-y] = \frac{\rho\,dx}{2\varepsilon_0}[(y^2+b^2-2yx)^{\frac{1}{2}}+x-y].$$

Integrating with respect to x over the range $0 \leqslant x \leqslant b$, this gives a net potential at P of

$$V = \frac{\rho}{12\varepsilon_0 y}[2(b^2+y^2)^{\frac{3}{2}} - 2y^3 - 3b^2y + 2b^3].$$

By expanding in ascending powers of $1/y$, it may be verified that $V \to 0$ as $y \to \infty$. At A, $y = b$, and $V = \rho b^2(4\sqrt{2}-3)/12\varepsilon_0$. Hence the loss in potential energy of the charge $-2\pi\rho b^3/3$ as it moves from a great distance to A is $\pi\rho^2 b^5(4\sqrt{2}-3)/18\varepsilon_0$. This must be compensated by an equal gain of kinetic energy $\frac{1}{2}mv^2$, where v is the velocity of arrival at A. It now follows that

$$v^2 = \frac{\pi\rho^2 b^5}{9m\varepsilon_0}(4\sqrt{2}-3). \qquad \square$$

***Problem* 1.8** An anode of a cathode ray tube is a plate bounded by concentric circles of radii a and $5a$. It is to be assumed uniformly charged and non-conducting. An electron is stationary at a great distance from the plate on its axis. It moves under the attraction of the plate, passing through the centre with velocity u. Calculate its velocity v when at a distance $2a\sqrt{6}$ from the plate.

Solution. It follows from equation (1.14) that the potential due to the anode at a point on its axis distant z from its centre is given by

$$V = \frac{\sigma}{2\varepsilon_0}[\sqrt{(z^2+25a^2)}-\sqrt{(z^2+a^2)}].$$

Thus, at $z = 2a\sqrt{6}$ and $z = 0$, the electric potentials are $a\sigma/\varepsilon_0$ and $2a\sigma/\varepsilon_0$ respectively. If $-e$ is the charge on the electron, its potential energies at these two points are accordingly $-ae\sigma/\varepsilon_0$ and $-2ae\sigma/\varepsilon_0$. Since the initial total energy of the electron is zero, conservation of energy implies that

$$\tfrac{1}{2}mv^2 - \frac{ae\sigma}{\varepsilon_0} = \tfrac{1}{2}mu^2 - \frac{2ae\sigma}{\varepsilon_0} = 0,$$

m being the electron's mass. It follows immediately that $v = u/\sqrt{2}$. $\square$

1.3 Gauss's Law If S is a closed surface described in an electrostatic field and E_n is the component of $\boldsymbol{E}$ in the direction of the outward normal at any point on S, the surface integral of E_n over S is termed the *flux* of $\boldsymbol{E}$ out of S. Denoting the flux by N, we have

$$N = \int_S E_n dS = \int_S \boldsymbol{E}\,.\,\boldsymbol{n}dS, \qquad (1.15)$$

where $\boldsymbol{n}$ is the unit outwardly directed normal to S.

Gauss's Law follows from the inverse square law (L. Marder, *Vector Fields*, p. 26) (equation (1.2)) and states that

$$N = Q/\varepsilon_0 \qquad (1.16)$$

where Q is the total charge enclosed by S.

By applying this result to a sphere of radius r which is concentric with a uniformly charged spherical surface, it is easily shown that the field intensity of such a charge distribution is given by

$$E = \frac{Q}{4\pi\varepsilon_0 r^2}, \quad r > a; \qquad E = 0, \quad r < a.$$

Here, Q is the total charge and $\boldsymbol{E}$ is clearly in the radial direction. Thus, the external field is identical with that which would be generated if the whole charge Q were placed at the centre.

The corresponding potential function outside the sphere $r = a$ is

$$V = \frac{Q}{4\pi\varepsilon_0 r}$$

(see equation (1.6)). Inside this sphere, V must be constant and equal to its value at the surface, viz. $Q/(4\pi\varepsilon_0 a)$.

***Problem* 1.9** Charge is distributed uniformly throughout the interior of a sphere of radius a. Show that the magnitude of the field intensity within the sphere is proportional to the distance from the centre.

Solution. Considerations of symmetry indicate that $\boldsymbol{E}$ is radial. Take a Gauss surface in the form of a concentric sphere of radius $r(< a)$. If Q is the total charge, the charge contained by the Gauss sphere is Qr^3/a^3. The magnitude of $\boldsymbol{E}$ must be constant over this sphere and its flux out of the sphere is accordingly $4\pi r^2 E$. Gauss's law now yields $4\pi r^2 E = Qr^3/\varepsilon_0 a^3$. Thus $E = Qr/(4\pi\varepsilon_0 a^3)$. □

***Problem* 1.10** A distribution of electric charge lies within an infinite circular cylinder of radius a. The charge density at a distance $r(< a)$ from the axis is given by $\rho = 3Q(a-r)/\pi a^3$. Calculate the potential function.

Solution. It follows from the axial symmetry that the intensity at any point P distant r from the axis of the cylinder will be radial (Fig. 1.5). Construct a coaxial cylinder of unit length with its curved surface containing P. If $r < a$, the net charge contained in this cylinder is

$$\int_0^r 2\pi\rho r\,dr = \frac{6Q}{a^3}\int_0^r r(a-r)\,dr = \frac{Qr^2}{a^3}(3a-2r).$$

The flux of $\boldsymbol{E}$ out of the circular ends of the cylinder is zero and the flux out of the curved surface is $2\pi rE$. Hence, by Gauss's law, $2\pi rE = Qr^2(3a-2r)/(\varepsilon_0 a^3)$. This gives $E = Qr(3a-2r)/(2\pi\varepsilon_0 a^3)$.

If $r > a$, the total charge contained in the Gauss cylinder is Q and, hence, $E = Q/(2\pi\varepsilon_0 r)$.

If V is the potential, $E = -dV/dr$. Integration accordingly leads to the results

$$V = \frac{Q}{12\pi\varepsilon_0 a^3}(5a^3 - 9ar^2 + 4r^3) \qquad r \leqslant a,$$

$$= -\frac{Q}{2\pi\varepsilon_0}\log(r/a) \qquad r \geqslant a.$$

(*Note:* The datum point has been taken on the surface $r = a$, so that V vanishes on this surface.) □

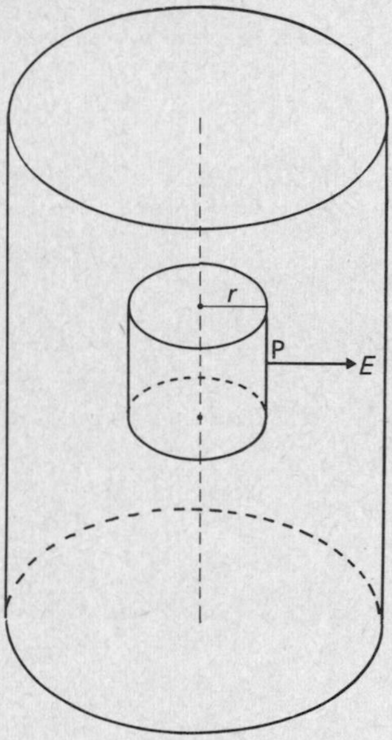

Fig. 1.5

The molecules of a conducting material contain charges which are free to move from molecule to molecule under the influence of any applied electric field. Thus, when a solid conductor is placed in an electric field, electric currents will commence to flow in its interior and will only cease when the electric intensity over the interior has been reduced to zero by a redistribution of charge (the *induced charge*). Under static conditions, therefore, $\boldsymbol{E} = 0$ within a conductor; a direct consequence of this is that V is constant over the conductor. In particular, the earth is an immense conductor whose potential is usually taken to be the datum with respect to which all other potentials are measured. Thus, the earth's potential is zero and when a conductor is 'earthed' its potential is reduced to zero.

Another consequence is that the interior of a conductor is uncharged; this follows since the flux out of any closed surface drawn inside the conductor is zero, and hence, by Gauss's law, it can never contain any charge.

1.4 Laplace's Equation Let S be any closed surface drawn in an electrostatic field and let Γ be the region enclosed by S. If ρ is the charge density, the total charge inside S is given by the volume integral $\int_\Gamma \rho dv$. Hence, by Gauss's law,

$$\int_S E_n \, dS = \varepsilon_0^{-1} \int_\Gamma \rho \, dv.$$

But, by the divergence theorem (L. Marder, *Vector Fields*, p. 43),

$$\int_S E_n \, dS = \int_\Gamma \operatorname{div} \boldsymbol{E} \, dv.$$

It now follows that

$$\int_\Gamma (\operatorname{div} \boldsymbol{E} - \rho/\varepsilon_0) \, dv = 0.$$

Since this is true for every region Γ, the integrand must vanish. Thus,

$$\operatorname{div} \boldsymbol{E} = \rho/\varepsilon_0. \tag{1.17}$$

This is the second fundamental equation of the electrostatic field.

Eliminating $\boldsymbol{E}$ between equation (1.7) and (1.17), we obtain Poisson's equation for the potential, namely,

$$\operatorname{div} \operatorname{grad} V = \nabla^2 V = -\rho/\varepsilon_0. \tag{1.18}$$

Thus, at all points where there is no charge, V satisfies Laplace's equation,

$$\nabla^2 V = 0. \tag{1.19}$$

Any solution of this equation is called a *harmonic function.*

It is assumed in equation (1.17) that ρ is finite and continuous. If the charge is distributed over a surface S as a film having infinitesimal thickness (e.g. S may be the surface of a conductor), this will not be the case. In these circumstances, let the region on one side of S be called region 1 and that on the other side be region 2 and let $\boldsymbol{n}$ be the unit normal to S in the sense from 1 to 2 (Fig. 1.6). Then, if E_{1n}, E_{2n} are the components of $\boldsymbol{E}$ in the direction of $\boldsymbol{n}$ on the two sides of S, the flux of $\boldsymbol{E}$ out of the small cylinder shown in the figure is $(E_{2n} - E_{1n})\omega$, where ω is the area of its right-section

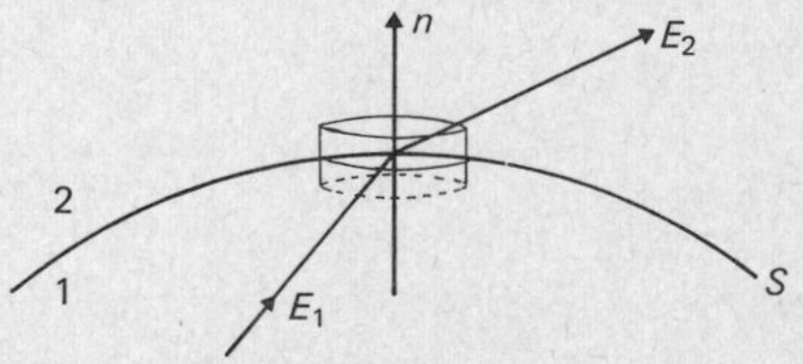

Fig. 1.6

(the curved surface of the cylinder makes negligible contribution if the length of the cylinder tends to zero). If σ is the surface density of the charge on S, the charge enclosed by the cylinder is $\sigma\omega$. Thus, by Gauss's law,

$$E_{2n}-E_{1n} = \sigma/\varepsilon_0. \tag{1.20}$$

In particular, if S is the surface of a conductor and region 1 is its interior, then $\boldsymbol{E}_1 = 0$ and equation (1.20) reduces to

$$E_n = \sigma/\varepsilon_0. \tag{1.21}$$

(The subscript 2 can now be dropped.)

It is further assumed that the potential V is continuous across a surface of discontinuity. This is equivalent to the statement that any component of $\boldsymbol{E}$ tangential to the surface is continuous. Thus, if the surface is that of a conductor, any tangential component of $\boldsymbol{E}$ just outside the conductor is zero and $\boldsymbol{E}$ is accordingly normal to the surface everywhere; this also follows from the fact that such a surface is an equipotential.

For use in many of the later problems, it will be convenient to list a number of well-known harmonic functions.

If (ρ, ϕ, z) are cylindrical polar coordinates (Fig. 1.7) the following are solutions of $\nabla^2 V = 0$:

$$V = A\log\rho + B, \tag{1.22}$$

$$V = (A\rho^n + B\rho^{-n})(C\cos n\phi + D\sin n\phi), \tag{1.23}$$

where A, B, C, D are arbitrary constants and n is an integer. Since these solutions are independent of z, they give the same field in every plane parallel to Oxy.

If (r, θ, ϕ) are spherical polar coordinates (Fig. 1.7), the following is a solution of Laplace's equation

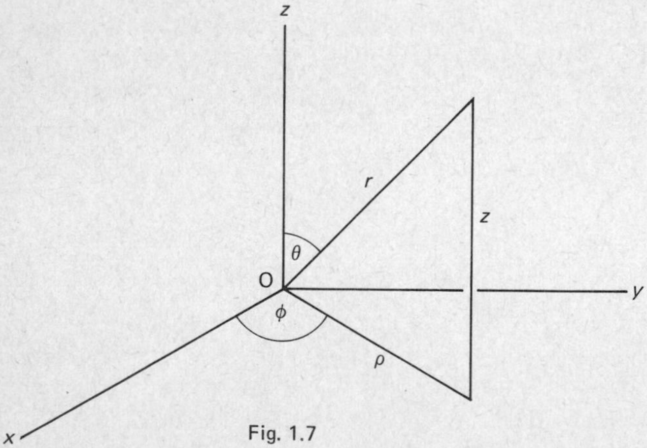

Fig. 1.7

$$V = \left(Ar^n + \frac{B}{r^{n+1}}\right)P_n(\cos\theta), \tag{1.24}$$

where n is zero or a positive integer and $P_n(\mu)$ is *Legendre's polynomial* of degree n; in particular

$$\begin{aligned} P_0(\mu) &= 1, \\ P_1(\mu) &= \mu, \\ P_2(\mu) &= \tfrac{1}{2}(3\mu^2 - 1), \\ P_3(\mu) &= \tfrac{1}{2}(5\mu^3 - 3\mu). \end{aligned} \tag{1.25}$$

More complex solutions can be derived from equations (1.22)–(1.24) by addition (*Principle of Superposition*).

***Problem* 1.11** A solid conducting sphere of radius a is surrounded by a concentric conducting spherical shell of internal radius b, which is earthed. If the inner sphere is raised to a potential ϕ, calculate its total charge.

Solution. Let P be any point in the field between the spheres and let OP $= r$ (Fig. 1.8). In this region, V must satisfy Laplace's equation and be

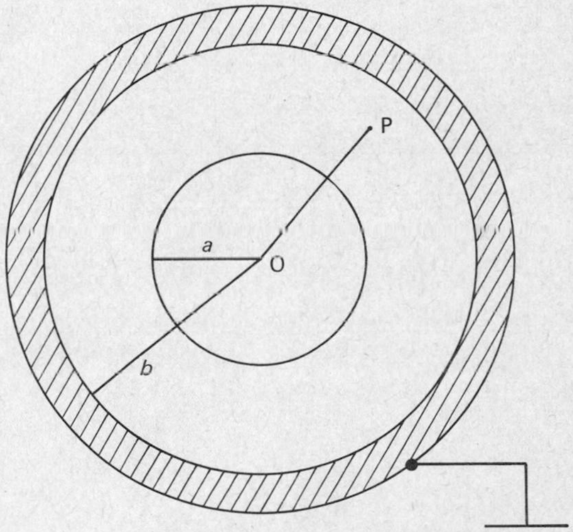

Fig. 1.8

spherically symmetric about O. With $n = 0$, the solution (1.24) is independent of θ and, hence, spherically symmetric. Thus, we take

$$V = A + \frac{B}{r}. \tag{1.26}$$

V has to satisfy the *boundary conditions* (i) $V = \phi$ over $r = a$, (ii) $V = 0$ over $r = b$. Thus, A and B have to be chosen such that

$$A + \frac{B}{a} = \phi, \qquad A + \frac{B}{b} = 0.$$

Solving for A, B and substituting in equation (1.26), we find

$$V = \frac{a\phi}{b-a}\left(\frac{b}{r} - 1\right). \tag{1.27}$$

Clearly, the electric intensity is directed radially and its magnitude is accordingly given by

$$E = -\frac{\partial V}{\partial r} = \frac{ab\phi}{b-a} \cdot \frac{1}{r^2}. \tag{1.28}$$

Thus, the flux of $\boldsymbol{E}$ across a concentric sphere of radius r is

$$4\pi r^2 E = \frac{4\pi ab\phi}{b-a}.$$

By Gauss's law, this equals Q/ε_0, where Q is the charge on the insulated sphere. Hence

$$Q = \frac{4\pi\varepsilon_0\, ab\phi}{b-a}. \tag{1.29}$$

Alternatively, the surface density of the charge on the sphere $r = a$ can be calculated from equation (1.21) to be

$$\sigma = \varepsilon_0 E_{r=a} = \frac{\varepsilon_0 b\phi}{a(b-a)} \tag{1.30}$$

and this then yields Q as before. □

***Problem* 1.12** An earthed conducting sphere of radius a is placed in a uniform field of intensity E_0. Calculate the charge induced on a hemisphere whose axis is in the direction of the field.

Solution. Taking the centre of the sphere as origin O and the z-axis in the direction of the uniform field E_0 (Fig. 1.9), let r, θ be spherical polar coordinates of any point P of the field (the third coordinate ϕ will not be needed, since the field has axial symmetry about Oz). In the absence of the sphere, the uniform field has potential V satisfying the equations $\partial V/\partial x = \partial V/\partial y = 0$, $\partial V/\partial z = -E_0$; thus

$$V = -E_0 z = -E_0 r\cos\theta = -E_0 r P_1(\cos\theta).$$

When the sphere is introduced, the charge induced on its surface causes a distortion of the uniform field, but this distortion will vanish at great

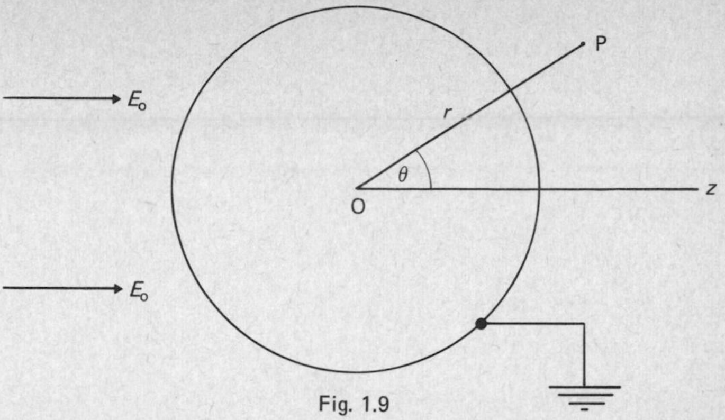

Fig. 1.9

distances from the sphere. Referring to equation (1.24), taking $n = 1$, we note that $BP_1(\cos\theta)/r^2$ is a solution of Laplace's equation vanishing at infinity. Assuming this to be the potential due to the induced charge, the resultant potential of the field becomes

$$V = -E_0 r \cos\theta + \frac{B}{r^2}\cos\theta. \tag{1.31}$$

It remains to satisfy the boundary condition over the sphere by appropriate choice of B. This condition is $V = 0$ over $r = a$ and requires that $B = E_0 a^3$. Hence

$$V = E_0\left(\frac{a^3}{r^3} - 1\right) r \cos\theta. \tag{1.32}$$

It has been shown that this potential function satisfies all the given conditions of the problem, but it is by no means self-evident that other potentials do not exist which also satisfy all these conditions. However, a *uniqueness theorem* can be proved (L. Marder, *Vector Fields*, p. 55) which asserts that this is not the case and equation (1.32) is accordingly established as the proper solution to our problem. Such uniqueness theorems will be very frequently appealed to in similar circumstances throughout the remainder of this book. On the surface of a conductor the conditions to be satisfied for uniqueness are that either the potential must take a given value or the total charge must be as given; if the charge is given, a further condition is that the potential must be constant (unknown) over the conductor.

The radial component of $\boldsymbol{E}$ can now be calculated from equation (1.32), thus,

$$E_r = -\frac{\partial V}{\partial r} = E_0\left(\frac{2a^3}{r^3} + 1\right)\cos\theta. \tag{1.33}$$

Putting $r = a$, this gives the normal component of $\boldsymbol{E}$ over the sphere to be $3E_0 \cos\theta$. Equation (1.21) then shows that the surface density of the induced charge is $\sigma = 3\varepsilon_0 E_0 \cos\theta$, i.e. the induced charge is positive over the hemisphere $0 < \theta < \frac{1}{2}\pi$ and is negative over the other hemisphere. By dividing the positively charged hemisphere into elemental rings by planes perpendicular to Oz, the total charge on this hemisphere is proved to be

$$2\pi a^2 \int_0^{\frac{1}{2}\pi} \sigma \sin\theta \, d\theta = 3\pi a^2 \varepsilon_0 E_0. \qquad \square \qquad (1.34)$$

***Problem* 1.13** A dipole of moment m is placed at the centre of a spherical cavity of radius a in an earthed conductor. Calculate the field.

Solution. Introducing spherical polar coordinates r, θ with pole O at the centre of the cavity and z-axis along the axis of the dipole (Fig. 1.10),

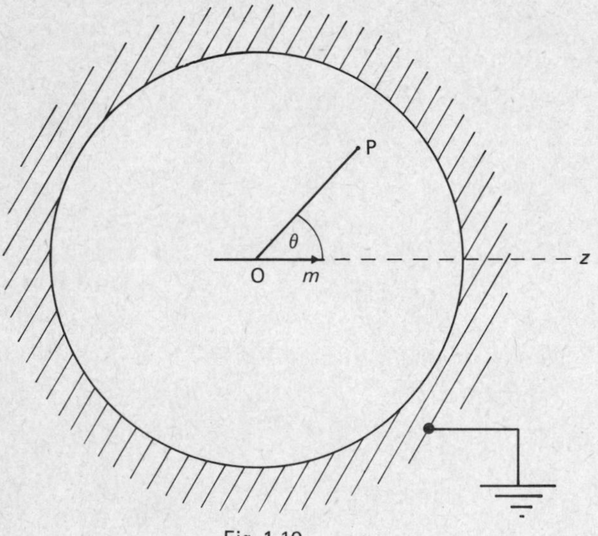

Fig. 1.10

we know from equation (1.9) that the potential due to the dipole alone is $m \cos\theta / 4\pi\varepsilon_0 r^2$. To this potential function must be added a term representing the contribution of the charge induced on the wall of the cavity. This term must have axial symmetry about Oz and must be finite within the cavity; we shall assume it to take the form $ArP_1(\cos\theta) = Ar\cos\theta$ (the case $B = 0$, $n = 1$, of equation (1.24)). Thus, the total potential within the cavity is given by

$$V = \frac{m\cos\theta}{4\pi\varepsilon_0 r^2} + Ar\cos\theta. \qquad (1.35)$$

It remains to satisfy the boundary condition $V = 0$ over $r = a$. Equation (1.35) shows that V vanishes for all values of θ when $r = a$, provided $A = -m/4\pi\varepsilon_0 a^3$. Thus,

$$V = \frac{m}{4\pi\varepsilon_0 a^2}\left(\frac{a^2}{r^2} - \frac{r}{a}\right)\cos\theta. \tag{1.36}$$

The radial component of the field is now calculated to be

$$E_r = -\frac{\partial V}{\partial r} = \frac{m}{4\pi\varepsilon_0 a^3}\left(\frac{2a^3}{r^3} + 1\right)\cos\theta. \tag{1.37}$$

Putting $r = a$, we find that the intensity at the wall of the cavity is $-3m\cos\theta/4\pi\varepsilon_0 a^3$. Thus, the surface density of the charge induced on the wall is given by $\sigma = -3m\cos\theta/4\pi a^3$. □

1.5 Force on the Surface of a Conductor The film of charge on the surface of a conductor is situated in a field whose intensity decreases very rapidly across the film from σ/ε_0 just outside the conductor to 0 inside the conductor. The average field is accordingly $\sigma/2\varepsilon_0$ and the force per unit area applied to the conductor surface can be proved to be $\sigma^2/2\varepsilon_0$ directed along the outwards normal.

***Problem* 1.14** A large plane conducting plate of area A is raised to potential ϕ and is placed at a distance d from a similar parallel plate which is earthed. Neglecting edge effects, show that it is attracted to the earthed plate with a force $\varepsilon_0 A\phi^2/2d$.

Solution. Taking the origin O on the earthed plate and the x-axis perpendicular to this plate (Fig. 1.11), if we neglect distortion in the region of the edges, the potential V will depend upon x alone. Thus, between the plates, V will satisfy $\nabla^2 V = d^2V/dx^2 = 0$ and, hence, $V = A + Bx$. Since

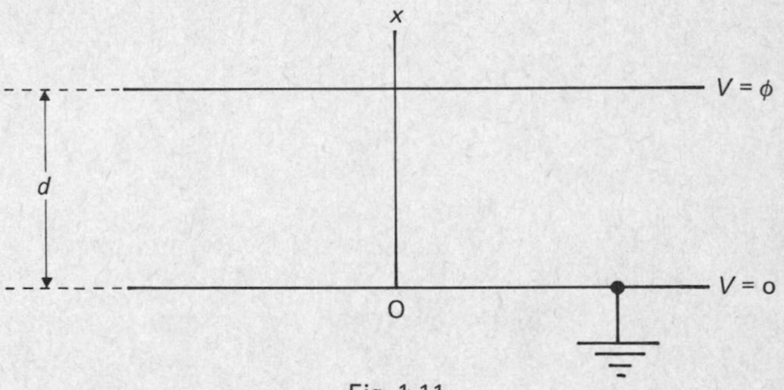

Fig. 1.11

$V = 0$ over $x = 0$, $V = \phi$ over $x = d$, these boundary conditions require that $A = 0$, $B = \phi/d$. It follows that

$$V = \phi x/d. \tag{1.38}$$

This result shows that $\boldsymbol{E}$ is parallel to Ox and that its magnitude is $E = -dV/dx = -\phi/d$. It follows from equation (1.21) that the surface density of the charge on the side of the insulated plate facing the earthed plate is given by $\sigma = \varepsilon_0 \phi/d$ and this corresponds to a force per unit area of $\sigma^2/2\varepsilon_0 = \varepsilon_0 \phi^2/2d$. Thus, the total force applied to the insulated plate is $\varepsilon_0 A\phi^2/2d$. □

***Problem* 1.15** A conducting sphere is cut into two hemispheres, earthed, and placed in a uniform field with the cut normal to the field. Calculate the force tending to separate the hemispheres.

Solution. The field has been calculated in Problem 1.12. The surface density of charge induced on the sphere was shown to be $3\varepsilon_0 E_0 \cos\theta$. Thus the force per unit area applied to the surface is $\frac{9}{2}\varepsilon_0 E_0^2\cos^2\theta$.

Divide one of the hemispherical surfaces into ring-shaped elements by planes perpendicular to its axis of symmetry as indicated in Fig. 1.12.

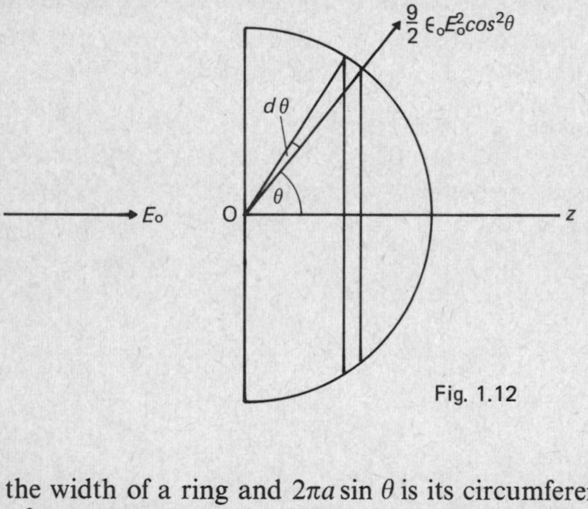

Fig. 1.12

Then $a\,d\theta$ is the width of a ring and $2\pi a\sin\theta$ is its circumference. Hence, its area is $2\pi a^2\sin\theta\,d\theta$. The resultant force on the ring must be along the axis of symmetry $\theta = 0$ and therefore only the component of the pressure $\frac{9}{2}\varepsilon_0 E_0^2\cos^2\theta$ in this direction, namely $\frac{9}{2}\varepsilon_0 E_0^2\cos^3\theta$, contributes. Thus, the resultant force on the ring is $9\varepsilon_0 \pi a^2 E_0^2\cos^3\theta\sin\theta\,d\theta$. Integrating for all rings belonging to the hemisphere, we obtain for the force tending to separate the hemispheres

$$\int_0^{\frac{1}{2}\pi} 9\varepsilon_0\, \pi a^2 E_0^2 \cos^3\theta \sin\theta\, d\theta = \tfrac{9}{4}\varepsilon_0\, E_0^2 \pi a^2. \qquad \square$$

***Problem* 1.16** The centre of the inner sphere of the system described in Problem 1.11 is displaced a small distance δ. Calculate the resultant force on this sphere to the first order in δ.

Solution. Let O, O′ be the centres of the inner and outer spheres respectively. Let (r, θ) be spherical polar coordinates relative to O as pole and OO′ as reference line. In the region between the spheres, we assume that the potential is given by

$$V = A + \frac{B}{r} + \left(Cr + \frac{D}{r^2}\right)\cos\theta. \tag{1.39}$$

When $\delta = 0$, this must reduce to the form of equation (1.26); C and D must accordingly be of order δ. Over $r = a$, $V = \phi$ for all θ; this boundary condition leads to the equations $A + B/a = \phi$, $Ca + D/a^2 = 0$. On the outer sphere, $r = b + \delta\cos\theta$ to the first order in δ and V must vanish. Substituting in equation (1.39) and approximating to the first order in δ, we derive the condition

$$A + \frac{B}{b}\left(1 - \frac{\delta}{b}\cos\theta\right) + \left(Cb + \frac{D}{b^2}\right)\cos\theta = 0.$$

This is satisfied for all θ provided $A + B/b = 0$, $Cb + D/b^2 = \delta B/b^2$. A, B, C, D can now be found. A and B have the values already calculated in Problem 1.11 and $a^3 C = -D = (\delta a^4 b\phi)/[(b-a)(b^3-a^3)]$.

The charge density on the inner sphere is next calculated to be given by

$$\sigma = -\varepsilon_0\left(\frac{\partial V}{\partial r}\right)_{r=a} = \varepsilon_0\left[\frac{B}{a^2} + \left(\frac{2D}{a^3} - C\right)\cos\theta\right].$$

The resultant force on the sphere is now determined as in the previous problem to be

$$\int_0^{\pi} \frac{\sigma^2}{2\varepsilon_0}\cos\theta \,.\, 2\pi a^2 \sin\theta\, d\theta = 2\pi\varepsilon_0 B\left(\frac{2D}{a^3} - C\right)\int_0^{\pi} \cos^2\theta \sin\theta\, d\theta$$

$$= \frac{4\pi\varepsilon_0\, \delta a^2 b^2 \phi^2}{(b-a)^2(b^3-a^3)}. \qquad \square$$

1.6 Images In some problems it is possible to imagine that the field is generated by point charges alone and to show that this field satisfies all the boundary conditions. Some of the point charges used will have no real existence and are called *images*; nevertheless, their combined potential will satisfy Laplace's equation and, provided it also satisfies the necessary boundary conditions, it follows from the uniqueness theorem that it represents the correct solution to the problem.

A point charge q placed at a point A distant d from an infinite earthed conducting plane provides a simple example. The potential generated by this charge and an image charge $-q$ situated at the optical image A′ of A in the plane (i) vanishes on the plane, (ii) vanishes at infinity, (iii) possesses the appropriate singularity in the neighbourhood of the charge at A. By the uniqueness theorem, it is accordingly the correct potential for the field between the charge and the plane.

***Problem* 1.17** Two semi-infinite, earthed, conducting planes meet at right angles at a common edge. A charge q is placed at a point A on a bisector of the angle between the planes at a distance d from the common edge. Show that the charge is attracted towards this edge with a force $(2\sqrt{2}-1)q^2/(16\pi\varepsilon_0 d^2)$.

Solution. The image system for the field in the angle between the planes comprises charges $-q$, q, $-q$ at points B, C, D respectively, where ABCD is a square (Fig. 1.13). This system of charges clearly generates zero potential over both planes as required. Thus, if the planes were removed and the image charges placed in position, the field in the neighbourhood

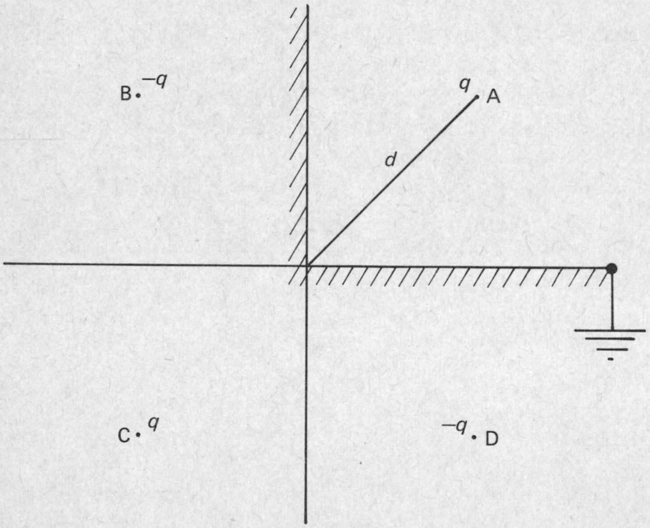

Fig. 1.13

of A would be unaltered. It follows that the force on the charge at A would be unchanged. This force can now be easily calculated by application of Coulomb's law and the result stated derived. □

***Problem* 1.18** A point charge q is placed outside an insulated, uncharged sphere of radius a at a distance f from the centre. Calculate the surface density of the charge on the sphere at a point distant r from q.

Solution. The image system comprises a charge $-aq/f$ at the point A′ inverse to A in the sphere and a charge aq/f at the centre O (Fig. 1.14). For the potential generated by the three charges at a point P on the sphere is

$$\frac{1}{4\pi\varepsilon_0}\left(\frac{q}{r}-\frac{aq}{f}\cdot\frac{1}{r'}+\frac{aq}{f}\cdot\frac{1}{a}\right). \tag{1.40}$$

Now, OA . OA′ = OP^2; hence, the triangles OAP, OPA′ have the angle at O in common and the sides about this angle in the same ratio, i.e.

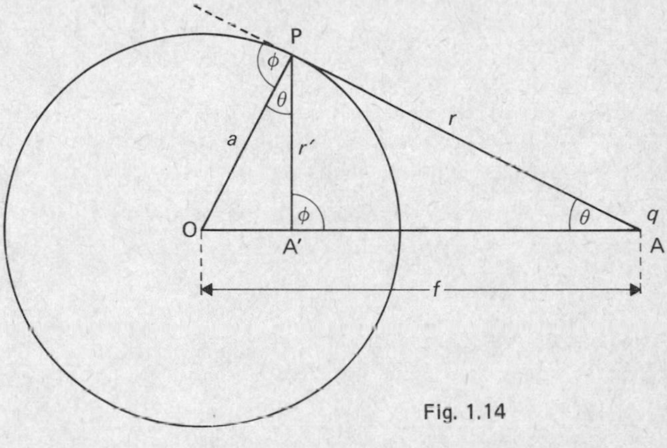

Fig. 1.14

OA/OP = OP/OA′. These triangles are similar, therefore, and this implies that $r/r' = f/a$. The expression (1.40) accordingly reduces to $q/(4\pi\varepsilon_0 f)$, showing that the three charges yield a constant potential over the sphere. Further, since the image charges enclosed by the sphere cancel, it follows from Gauss's law that the flux of electric intensity out of the sphere generated by the three charges is zero. But, if σ is the surface density of charge on the sphere S, then

$$\text{charge on sphere} = \int_S \sigma\, dS = \varepsilon_0 \int_S E_n\, dS = 0.$$

Thus, all the boundary conditions are satisfied by the image system.

Resolving the intensities at P due to the charges along OP, we find that

$$E_n = \frac{1}{4\pi\varepsilon_0}\left[\frac{aq}{f}\cdot\frac{1}{a^2} - \frac{aq}{f}\cdot\frac{1}{r'^2}\cos\theta - \frac{q}{r^2}\cos\phi\right]. \tag{1.41}$$

Substituting $r' = ar/f$, $\cos\theta = (r^2 + f^2 - a^2)/2rf$, $\cos\phi = (f^2 - a^2 - r^2)/2ar$ and noting that $\sigma = \varepsilon_0 E_n$, equation (1.41) gives

$$\sigma = \frac{q}{4\pi\varepsilon_0}\left[\frac{1}{af} - \frac{f^2 - a^2}{ar^3}\right].$$ □

***Problem* 1.19** A dipole of moment m is placed at a distance f from the centre of an earthed sphere of radius a. The axis of the dipole points directly away from the sphere. Show that the charge induced is am/f^2.

Solution. Suppose the charges $-q$, q of the dipole lie at the points A, B respectively (Fig. 1.15), where OA $= f$, OB $= f + \delta f$. Then $m = q\,\delta f$ and $\delta f \to 0$. The inverse points A′, B′ are such that

$$\mathrm{OA}' = \frac{a^2}{f}, \qquad \mathrm{OB}' = \frac{a^2}{(f+\delta f)} = \frac{a^2}{f}\left(1 - \frac{\delta f}{f}\right)$$

to the first order in δf. Hence, A′B′ $= a^2 \delta f/f^2$. The sphere will be at zero potential if we introduce image charges aq/f at A′ and $-aq/(f+\delta f)$ at B′. To the first order in δf, $-aq/(f+\delta f) = -(aq/f) + (aq/f^2)\delta f$. Thus, the charges at A′, B′ are equivalent to a dipole of moment

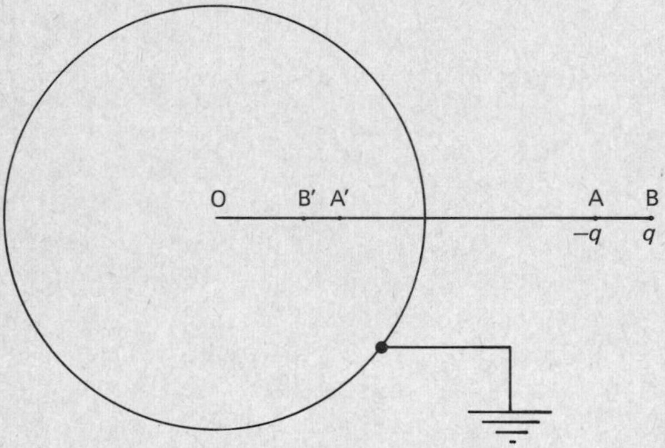

Fig. 1.15

$aq\,.\,\mathrm{A'B'}/f = a^3 q\,\delta f/f^3 = a^3 m/f^3$ directed along B′A′ and a further charge $aq\,\delta f/f^2 = am/f^2$ at B′. In the limit, as $\delta f \to 0$, both the image dipole $a^3 m/f^3$ and the image charge am/f^2 are situated at A′.

The charge induced on the sphere is the total image charge enclosed, i.e. am/f^2. □

***Problem* 1.20** A closed hemispherical shell of radius a is constructed from sheet metal and is earthed. A charge q is placed inside the shell, on its axis, a distance f from its centre. Find the force experienced by the charge.

Solution. If q is placed at A, image charges $-aq/f$, $-q$, aq/f are introduced at the points A′, B, B′ respectively (Fig. 1.16). A, A′ are inverse

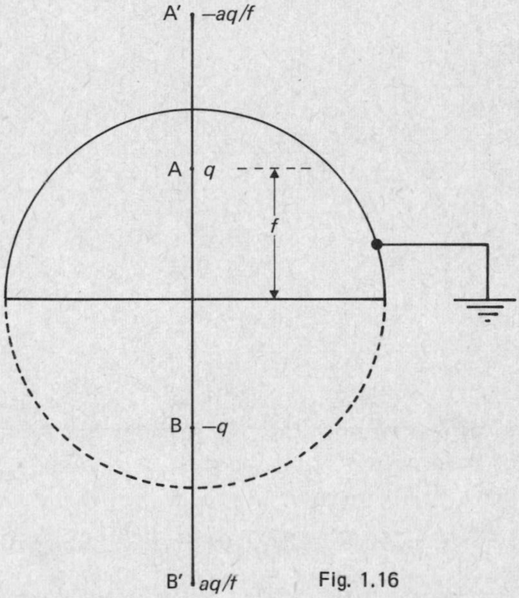

Fig. 1.16

points with respect to the completed sphere; B is the optical image of A in the plane face of the hemisphere; B′ is the inverse of B with respect to the sphere. Clearly, B′ is the optical image of A′ in the plane face. The charges at A and A′ and the charges at B and B′ make the potential on the curved surface of the hemisphere vanish. Also, the charges at A and B and the charges at A′ and B′ make the potential zero over the plane face. Thus, the image system solves the problem.

In particular, the resultant force on the original charge at A due to the images is

$$\frac{q^2}{4\pi\varepsilon_0}\left[\frac{1}{4f^2}-\frac{a/f}{(f+a^2/f)^2}-\frac{a/f}{(f-a^2/f)^2}\right]$$

directed towards B. □

***Problem* 1.21** An infinite line charge of density q per unit length is fixed parallel to the axis of an infinite circular conducting cylinder of radius a. The distance between the line charge and cylinder axis is f. Calculate the charge per unit length of the cylinder if no force is exerted on the line charge.

Solution. A right-section of the cylinder and line charge is shown in Fig. 1.17. The line charge and cylinder axis meet the plane of section in A

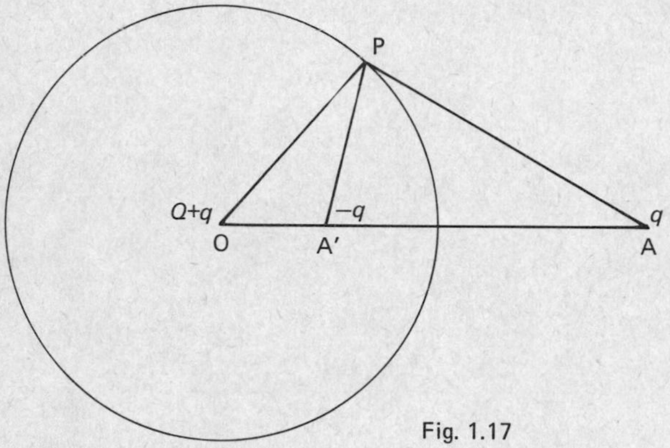

Fig. 1.17

and O respectively. Introduce parallel image line charges $-q$, $Q+q$ to meet the plane of section in A′ and O respectively, A′ being the point inverse to A in the circular section. Then, at any point P on the cylinder, the potential generated is given by

$$\begin{aligned} V &= -2q \log \mathrm{AP} + 2q \log \mathrm{A'P} - 2(Q+q) \log \mathrm{OP} \\ &= 2q \log\left(\frac{\mathrm{A'P}}{\mathrm{AP}}\right) - 2(Q+q) \log a \\ &= -2q \log f - 2Q \log a. \end{aligned}$$

This is constant over the whole surface of the cylinder as required. Since the total image charge per unit length enclosed by the cylinder is Q, this is its charge per unit length.

The field intensity at A due to the two images is

$$\frac{1}{4\pi\varepsilon_0}\left[\frac{2(Q+q)}{f} - \frac{2q}{f - a^2/f}\right].$$

This vanishes if $Q = a^2 q/(f^2 - a^2)$. □

***Problem* 1.22** A pair of infinite circular conducting cylinders, both of radius a, are placed with their axes parallel a distance $2d$ apart. Their potentials are $+\phi$ and $-\phi$. Calculate their charges.

Solution. A right-section of the cylinders by a plane meets their axes in O and O′ (Fig. 1.18). A, A′ are points on OO′ which are inverse with respect to both circular sections. Thus, if OA′ = O′A = x, then

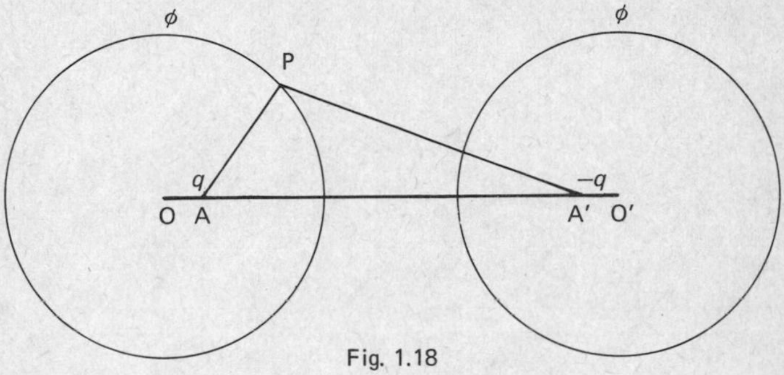

Fig. 1.18

OA = O′A′ = $2d-x$ and, hence, $x(2d-x) = a^2$. Solving this last equation for x, we find $x = d+\sqrt{(d^2-a^2)}$. (N.B. positive root since $x > d$.)

Introduce parallel infinite image line charges q, $-q$ to intersect the plane of section at A, A′ respectively. The potential generated at any point P on the left-hand cylinder is now given by

$$\phi = -\frac{q}{2\pi\varepsilon_0}\log \mathrm{AP}+\frac{q}{2\pi\varepsilon_0}\log \mathrm{A'P} = \frac{q}{2\pi\varepsilon_0}\log\left(\frac{x}{a}\right).$$

This is constant, as required. To give the correct potential, it is necessary that

$$q = \frac{2\pi\varepsilon_0\,\phi}{\log[d/a+\sqrt{(d^2/a^2-1)}]}.$$

The potential over the other cylinder can now be shown to be $-\phi$, similarly.

The last equation gives the charge per unit length over the left-hand cylinder. The charge on the other cylinder is $-q$ per unit length.

1.7 Capacitors. Energy An insulated conductor specially designed to store electric charge is called a condenser or a *capacitor*. Such a device usually comprises two plates, one being earthed and the other being insulated and serving to hold the charge. It is found that the charge Q

stored is directly proportional to the potential ϕ above earth of the insulated conductor; the ratio Q/ϕ is called the *capacitance*. The purpose of the earthed conductor is to increase the capacitance.

Making use of results stated in § 1.2, Problems 1.11 and 1.14, the capacitances of (a) a conducting sphere of radius a, (b) the same sphere surrounded by an earthed spherical shell of radius b, and (c) a parallel plate capacitor whose plates have area A and are separated by a distance d, are all easily found. The results are (a) $4\pi\varepsilon_0 a$, (b) $4\pi\varepsilon_0 ab/(b-a)$, (c) $\varepsilon_0 A/d$.

***Problem* 1.23** The axis of an infinite circular conducting cylinder of radius a is parallel to an infinite earthed plane and distant d from it. Find the capacitance per unit length of the cylinder.

Solution. The image line charges q, $-q$ at A and A′ in Fig. 1.18 have been shown (previous problem) to generate a constant potential ϕ over the left-hand cylinder, provided

$$q = \frac{2\pi\varepsilon_0 \phi}{\log[d/a+(d^2/a^2-1)^{\frac{1}{2}}]}.$$

They also generate zero potential over the plane which is the perpendicular bisector of OO′. This image system accordingly solves our present problem.

By Gauss's law, q will be the charge per unit length on the cylinder. Hence, its capacitance per unit length is

$$\frac{q}{\phi} = \frac{2\pi\varepsilon_0}{\log[d/a+(d^2/a^2-1)^{\frac{1}{2}}]}.$$ □

If charges $q_1, q_2, \ldots, q_n$ are situated at points $P_1, P_2, \ldots, P_n$ and V_i is the potential at P_i due to all charges except q_i, the potential energy U of the system is given by

$$U = \tfrac{1}{2}\sum_{i=1}^{n} q_i V_i. \tag{1.42}$$

This is the work which would be done by the attractions and repulsions between the charges if these were separated to be at great distances from one another.

In the case of a capacitor C whose insulated plate has been raised to potential ϕ, the charge on this plate is $C\phi$; the contribution of this charge to the energy of the system is therefore $\frac{1}{2}C\phi^2$. The earthed plate is at zero potential and the induced charge upon it accordingly makes zero contribution. Thus, the total energy of the capacitor is $\frac{1}{2}C\phi^2$.

***Problem* 1.24** Calculate the internal electrical energy of a proton assuming that its charge e is distributed uniformly over a sphere of radius a.

Solution. Employing the result found in the solution to Problem 1.9, the field intensity in the proton will be radial and of magnitude $er/(4\pi\varepsilon_0 a^3)$. Since $E = -dV/dr$, this implies that $V = C - er^2/(8\pi\varepsilon_0 a^3)$, where C is constant. Outside the proton, the field is the same as if all the charge were at the centre; thus, $V = e/(4\pi\varepsilon_0 r)$. Since V must be continuous over the surface $r = a$, $C = 3e/(8\pi\varepsilon_0 a)$.

Consider the charge enclosed between spheres of radii r, $r+dr$ with their centres at the centre of the proton. The quantity of charge is $3er^2 dr/a^3$ and its potential is $e(3a^2 - r^2)/(8\pi\varepsilon_0 a^3)$. Its contribution to the internal energy is therefore $3e^2r^2(3a^2-r^2)dr/(16\pi\varepsilon_0 a^6)$. Integrating over the interval $0 \leqslant r \leqslant a$, the total internal energy is found to be $3e^2/(20\pi\varepsilon_0 a)$.

***Problem* 1.25** A spherical conductor of radius a is kept by batteries at a constant potential ϕ. It is surrounded by a concentric, spherical conductor which is earthed. Under the internal forces this latter conductor contracts from radius b to b_1. Calculate the work done by these forces.

Solution. The charge on the inner sphere has already been calculated at equation (1.29). Thus, the energy of the system before contraction is $2\pi\varepsilon_0 ab\phi^2/(b-a)$.

After contraction, the energy is $2\pi\varepsilon_0 ab_1\phi^2/(b_1-a)$ and the gain in energy is accordingly $U = [2\pi\varepsilon_0 a^2(b-b_1)\phi^2]/[(b-a)(b_1-a)]$.

The increase in charge on the inner sphere is

$$\frac{4\pi\varepsilon_0 ab_1\phi}{b_1-a} - \frac{4\pi\varepsilon_0 ab\phi}{b-a} = \frac{4\pi\varepsilon_0 a^2(b-b_1)\phi}{(b-a)(b_1-a)}.$$

This charge is raised by the batteries from earth potential to potential ϕ. Thus, the work done by the batteries on the system is $2U$. Of this, one half appears as a gain in energy of the system and the other half corresponds to the work done by the internal forces. Hence, the work done by the internal forces is U as given above. □

EXERCISES

1. A line doublet comprises a pair of equal and opposite parallel infinite line charges a small distance d apart. If q and $-q$ are the charges per unit length, show, by neglecting powers of d after the first, that the potential of the field is given by $V = m\cos\phi/(2\pi\varepsilon_0\rho)$, where $m = qd$ is the *moment* of the line doublet and (ρ, ϕ) are cylindrical polar coordinates in a frame Oxyz for which Oz is parallel to the line charges and midway between them and Ox lies in the plane of these charges.

2. Prove that the field intensity due to a uniformly charged ring of total charge q and radius a, at a point on its axis distant z from its centre, is $qz(a^2+z^2)^{-\frac{3}{2}}/(4\pi\varepsilon_0)$.

A fixed cylindrical shell of radius a and length $2l$ carries a total charge Q uniformly distributed over it. A particle of mass m and charge $-q$ is free to perform small oscillations about the centre of the cylinder's axis, along the line of this axis. Prove that the period is $2\pi[4\pi\varepsilon_0 m(a^2+l^2)^{\frac{3}{2}}/Qq]$.

3. Infinite coaxial conducting cylinders have radii a and b ($a < b$). The outer cylinder is earthed and the inner is raised to a potential V_0. Assuming that the potential between the cylinders takes the form shown in equation (1.22), determine the field. Hence show that the capacitance per unit length of this cylindrical condenser is $2\pi\varepsilon_0/[\log(b/a)]$.

4. If the conducting sphere in Problem 1.12 is insulated and given a charge Q before being placed in the uniform field E_0, show that the potential of the resultant field outside the sphere is given by

$$V = E_0\left(\frac{a^3}{r^3}-1\right)r\cos\theta+\frac{Q}{4\pi\varepsilon_0 r}.$$

Deduce that the surface density of the charge on the sphere is positive everywhere provided $Q > 12\pi a^2\varepsilon_0 E_0$. (Assume $E_0 > 0$).

5. A conductor in the form of an infinite plane with a hemispherical boss of radius a, is earthed in the presence of a point charge q at a point P. The line OP makes an angle θ with the normal to the plane, where O is the centre of the circular base of the hemisphere and OP $= f$. Show that, if a/f is small, the component of attraction on the charge normal to the plane is greater than it would be in the absence of the boss by an amount of approximately $q^2a^3\cos\theta(1+3\cos 2\theta)/f^5$.

6. The centre of an insulated, uncharged conducting sphere of radius a is at the midpoint of the straight line joining two equal point charges, which are a distance $2f$ apart. If a/f is small, show that the force on either charge would be increased in the approximate ratio $(1+24a^5/f^5):1$, if the sphere were removed.

Chapter 2

Polarized Media

2.1 Poisson's Distribution When an insulator or dielectric is placed in an electric field it becomes *polarized*, i.e. any small element behaves like a dipole and contributes to the resultant field. This effect is due either to a slight separation of the positive and negative charges within the atoms or to the alignment of the axes of existing atomic dipoles under the influence of the field.

If dv is any small element of dielectric volume and $\boldsymbol{P}\,dv$ is its dipole moment, then $\boldsymbol{P}$ is called the *polarization* of the dielectric at the position of the element.

If a dielectric is bounded by a surface S, the field at any external point due to the polarization within S is identical with that due to a surface charge over S having density $\boldsymbol{P}.\boldsymbol{n}$ ($\boldsymbol{n}$ is the unit outwardly directed normal to S) and a charge distributed over the interior of S with volume density $-\operatorname{div}\boldsymbol{P}$. This is *Poisson's distribution.*

Inside a dielectric, $\boldsymbol{E}$ is *defined* to be the field intensity of the Poisson distribution. The actual force exerted upon unit charge situated inside a dielectric is difficult to define unambiguously. If the charge is supposed to be placed in a small cavity, the force depends on the shape of the cavity. If the cavity is needle-shaped, with its axis parallel to $\boldsymbol{P}$, the force can be shown to be $\boldsymbol{E}$ as just defined.

Problem 2.1 Calculate the field due to a uniformly polarized sphere.

Solution. Taking the centre of the sphere O as the origin for spherical polar coordinates $(r, \theta\,\phi)$, we shall suppose that the polarization $\boldsymbol{P}$ is always directed along the z-axis (Fig. 2.1).

Then, at any point (r, θ, ϕ) on the sphere's surface, $\boldsymbol{P}.\boldsymbol{n} = P\cos\theta$; this gives Poisson's surface distribution. $\operatorname{div}\boldsymbol{P} = 0$, so that there is no volume distribution.

The dielectric sphere can now be imagined to be removed and replaced by the surface distribution; the methods of Chapter 1 are then applicable. Let V_1 be the field potential within the sphere and V_2 the potential outside. Both potentials must satisfy Laplace's equation and we shall take

$$V_1 = Ar\cos\theta, \qquad V_2 = \frac{B}{r^2}\cos\theta.$$

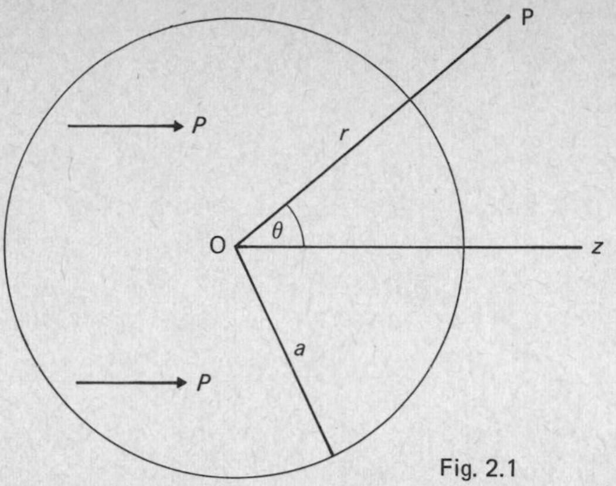

Fig. 2.1

The boundary conditions to be satisfied over $r = a$ are (i) V is continuous hence $Aa = B/a^2$ and (ii) $E_{n2} - E_{n1} = \sigma/\varepsilon_0 = P\cos\theta/\varepsilon_0$; since $E_n = -\partial V/\partial r$, the second condition yields $A + 2B/a^3 = P/\varepsilon_0$. Thus, $A = P/3\varepsilon_0$ and $B = Pa^3/3\varepsilon_0$.

These results imply that the internal field is uniform and of intensity $-\boldsymbol{P}/3\varepsilon_0$ and that the external field is identical with that due to a dipole of moment $4\pi \boldsymbol{P} a^3/3$ placed at O.

2.2 Fields in Dielectrics In most dielectrics, it is found that $\boldsymbol{P}$ is directly proportional to $\boldsymbol{E}$. Thus, we write

$$\boldsymbol{P} = \chi\varepsilon_0 \boldsymbol{E}, \tag{2.1}$$

where χ is called the *electric susceptibility*.

Replacing all polarized dielectrics by their Poisson distributions, the situation reverts to that studied in Chapter 1 and the field equations used in the absence of dielectrics become applicable. In particular,

$$\boldsymbol{E} = -\operatorname{grad} V. \tag{2.2}$$

Also, at any point there will be a density ρ of real charge and a density $-\operatorname{div} \boldsymbol{P}$ of imaginary Poisson charge. It follows from equation (1.17) that $\operatorname{div} \boldsymbol{E} = (\rho - \operatorname{div} \boldsymbol{P})/\varepsilon_0$. This equation can now be written in the form

$$\operatorname{div} \boldsymbol{D} = \rho, \tag{2.3}$$

where $\boldsymbol{D} = \varepsilon_0 \boldsymbol{E} + \boldsymbol{P} = \varepsilon_0(1+\chi)\boldsymbol{E}$, by equation (2.1). Writing $\varepsilon = 1+\chi$, this gives

$$\boldsymbol{D} = \varepsilon\varepsilon_0 \boldsymbol{E}. \tag{2.4}$$

ε is called the *dielectric constant*.

Equations (2.2)–(2.4) are the fundamental equations of the electrostatic field in the presence of dielectrics. $\boldsymbol{D}$ is called the *electric displacement*. If a small disc-shaped cavity is cut in a dielectric with its axis aligned with $\boldsymbol{P}$, the field intensity inside is $\boldsymbol{D}/\varepsilon_0$.

Problem 2.2 The potential for a field in a uniform dielectric whose constant is ε, is given by $V = V(r)$, where r is distance measured from a point O. Calculate the charge density ρ. If $\rho = \rho_0 (a/r)^2$, find V.

Solution. Equation (2.2) gives $\boldsymbol{E} = -V'\boldsymbol{r}/r$, where $V' = dV/dr$. By equation (2.4), $\boldsymbol{D} = -\varepsilon\varepsilon_0 V'\boldsymbol{r}/r$ and it then follows from equation (2.3) that

$$\rho = -\varepsilon\varepsilon_0 \operatorname{div}\left(\frac{V'}{r}\boldsymbol{r}\right) = -\varepsilon\varepsilon_0 (V''+2V'/r) = -\frac{\varepsilon\varepsilon_0}{r^2}\frac{d}{dr}(r^2V').$$

In the case $\rho = \rho_0(a/r)^2$, integrating the last equation we find that

$$V' = \frac{A}{r^2} - \frac{\rho_0 a^2}{\varepsilon\varepsilon_0 r},$$

where A is constant. It is easy to verify that the charge inside a small sphere of radius δ and centre O tends to zero with δ, i.e. there is no point charge at O. Hence $A = 0$. A further integration now gives

$$V = -\frac{a^2\rho_0}{\varepsilon\varepsilon_0}\log r + \text{constant}. \qquad \square$$

Equation (2.3) implies that the flux of $\boldsymbol{D}$ out of any closed surface S is equal to the total (actual) charge enclosed in S. This is the generalized form of Gauss's flux law (equation (1.16)).

Problem 2.3 Amend Coulomb's law (equation (1.3)) for the case when the two charges are immersed in an infinite dielectric of constant ε.

Solution. Consider the situation when the charge q is alone in the dielectric at a point O. The resulting field will be spherically symmetric about O and directed radially from this point. Thus, the flux of $\boldsymbol{D}$ out of a sphere of radius r and centre O is $4\pi r^2 D$. By Gauss's law, $4\pi r^2 D = q$. Hence, at a point P where OP $= \boldsymbol{r}$,

$$\boldsymbol{D} = \frac{q}{4\pi r^3}\boldsymbol{r}. \qquad (2.5)$$

Employing equation (2.4), this gives

$$\boldsymbol{E} = \frac{q}{4\pi\varepsilon\varepsilon_0 r^3}\boldsymbol{r}. \qquad (2.6)$$

Equation (2.2) then leads to

$$V = \frac{q}{4\pi\varepsilon\varepsilon_0 r}. \qquad (2.7)$$

Now introduce the charge q' at P (assumed placed in a radial needle-shaped cavity so that it is free to move in the direction of the applied force). The force exerted upon it is given by

$$\boldsymbol{F} = q'\boldsymbol{E} = \frac{qq'}{4\pi\varepsilon\varepsilon_0 r^3}\boldsymbol{r}. \tag{2.8}$$

This is the amended Coulomb's Law. □

2.3 Harmonic Functions Eliminating $\boldsymbol{E}$ and $\boldsymbol{D}$ between equations (2.2)–(2.4), we find that

$$\operatorname{div}(\varepsilon \operatorname{grad} V) = -\rho/\varepsilon_0 . \tag{2.9}$$

Over any region in which ε does not vary, this reduces to

$$\nabla^2 V = -\rho/\varepsilon\varepsilon_0 . \tag{2.10}$$

Thus, if ρ vanishes over such a region, V will be a harmonic function.

If a surface S carries a charge having density σ and the regions on either side of S are occupied by dielectrics of constants $\varepsilon_1, \varepsilon_2$, the following boundary condition is applicable:

$$D_{2n} - D_{1n} = \sigma. \tag{2.11}$$

This follows from Gauss's law by an argument similar to that used to derive equation (1.20). The subscript n indicates a component along the normal to S directed from the region 1 to the region 2; this is the amended form of the condition (1.20). If S is the surface of a conductor, then $\boldsymbol{D}_1 = 0$ inside the conductor and $\boldsymbol{D}_2$ is directed along the normal; thus, the condition reduces to

$$D_2 = \sigma. \tag{2.12}$$

The potential V is assumed to be continuous across S; this is equivalent to the statement that all tangenitial components of $\boldsymbol{E}$ are continuous.

Problem 2.4 Two conducting closed surfaces, one inside the other, form the plates of a capacitor. If the region between the surfaces is filled with uniform dielectric of constant ε, show that the capacitance is multiplied by ε.

Solution. Let S_0 be the outer surface and S_1 the inner surface. With no dielectric between the plates, suppose S_0 is earthed and S_1 is raised to potential ϕ. Then the potential V in the region between S_0 and S_1 is harmonic and satisfies the boundary conditions $V = 0$ on S_0 and $V = \phi$ on S_1. These conditions serve to determine V uniquely. When V has been found, the surface density of the charge on S_1 can be found from the equation $\sigma = \varepsilon_0 E_n = -\varepsilon_0\, \partial V/\partial n$, $\partial/\partial n$ denoting differentiation along the outward normal to S_1.

Now suppose the dielectric is introduced and the plate S_1 is maintained at the potential ϕ. Then V satisfies the same conditions as before and is therefore unchanged. However, the surface density of the charge on S_1 is now given by $\sigma' = D_n = -\varepsilon\varepsilon_0\,\partial V/\partial n = \varepsilon\sigma$, i.e. the total charge on S_1 is multiplied by ε. Since the potential difference between the plates is still ϕ, this means that the capacitance has been increased by a factor ε. □

Problem 2.5 A conducting sphere of radius a is coated with a dielectric shell of thickness d and constant ε. It is then surrounded by a concentric conducting shell of radius b, which is earthed. Find the capacitance of the inner sphere.

Solution. Suppose the inner conductor is raised to potential ϕ. In region 1 (Fig. 2.2), we assume $V = A+B/r$ and in region 2 we take $V = P+Q/r$. Over $r = a$, $V = \phi$ and over $r = b$, $V = 0$. Also, V is continuous across $r = (a+d)$. These conditions give $\phi = A+(B/a)$, $0 = P+(Q/b)$, $A+[B/(a+d)] = P+[Q/(a+d)]$.

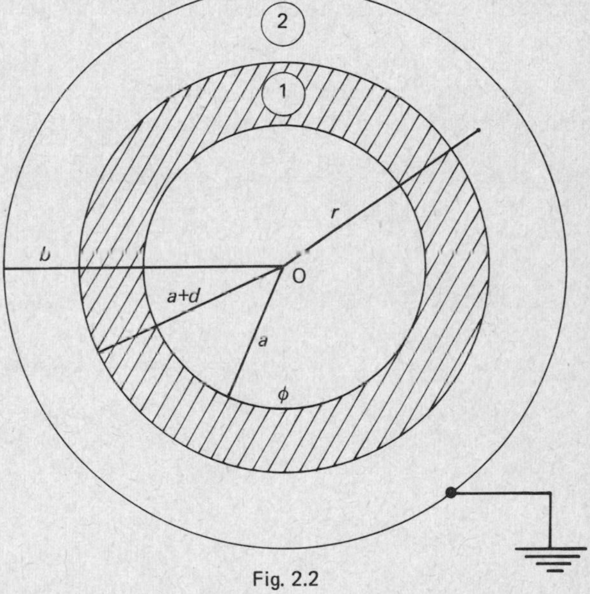

Fig. 2.2

In region 1, the field is radial and, hence, $E = -dV/dr = B/r^2$; thus, $D = \varepsilon\varepsilon_0\,B/r^2$. Similarly, in region 2, $D = \varepsilon_0 Q/r^2$. Applying condition (2.11) to the surface of discontinuity $r = a+d$, we get

$$\frac{\varepsilon_0 Q}{(a+d)^2} = \frac{\varepsilon\varepsilon_0 B}{(a+d)^2}.$$

We can now solve for A, B, P, Q thus:

$$A = \frac{a\varepsilon(b-a-d)-ab}{bd+a\varepsilon(b-a-d)}\phi, \qquad B = \frac{ab(a+d)}{bd+a\varepsilon(b-a-d)}\phi,$$

$$P = -\varepsilon B/b, \qquad Q = \varepsilon B.$$

Using equation (2.12), we calculate that the density of charge on the sphere $r = a$ is $\varepsilon\varepsilon_0 B/a^2$ and, hence, that the total charge is $4\pi\varepsilon\varepsilon_0 B$. Thus, the capacitance is

$$\frac{4\pi\varepsilon\varepsilon_0 B}{\phi} = \frac{4\pi\varepsilon\varepsilon_0 ab(a+d)}{bd+a\varepsilon(b-a-d)}.$$

A special case is when the dielectric fills the gap between the plates. Then, $d = b-a$ and the capacitance is $4\pi\varepsilon\varepsilon_0 ab/(b-a)$. Comparison of this result with that given in §1.7 reveals that the presence of the dielectric again increases the capacitance by a factor ε. □

Problem 2.6 A dielectric sphere of radius a and constant ε is placed in a uniform field of strength E_0. Find the resultant field produced.

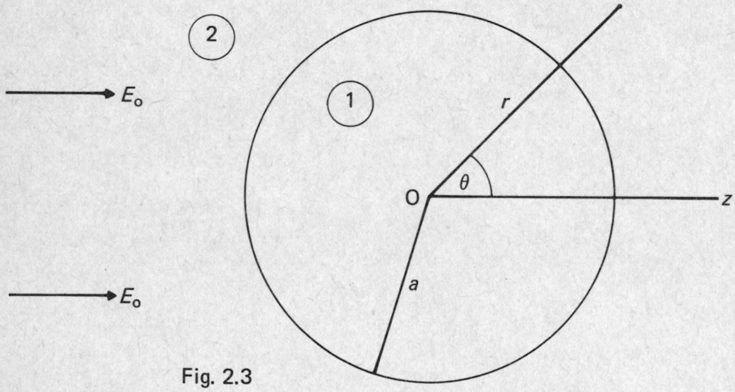

Fig. 2.3

Solution. Taking the origin of spherical polar coordinates (r, θ, ϕ) at the centre of the sphere and z-axis in the direction of the applied field (Fig. 2.3), in the absence of the sphere the potential is given by $V = -E_0 z = -E_0 r\cos\theta$. When the sphere is introduced, its polarization will distort the uniform field but the distortion will be negligible at great distances; accordingly we shall assume that, outside the sphere (region 2),

$$V = V_2 = -E_0 r\cos\theta + \frac{A}{r^2}\cos\theta.$$

This function is axially symmetric about Oz and satisfies Laplace's equation throughout region 2.

Inside the sphere (region 1), the potential must be axially symmetric and finite at $r = 0$. Hence, we assume $V = V_1 = Br\cos\theta$.

Across the surface $r = a$, both V and D_n must be continuous. Since $D_n = \varepsilon\varepsilon_0 E_n = -\varepsilon\varepsilon_0(dV/dr)_{r=a}$, these conditions are satisfied provided $-E_0 a + (A/a^2) = Ba$, $E_0 + (2A/a^3) = -\varepsilon B$. Thus, $A = a^3 E_0(\varepsilon - 1)/(\varepsilon + 2)$ and $B = -3E_0/(\varepsilon + 2)$. These give

$$V_1 = -\frac{3E_0 r\cos\theta}{\varepsilon + 2}, \qquad V_2 = -E_0 r\cos\theta + \frac{\varepsilon - 1}{\varepsilon + 2}\cdot\frac{a^3 E_0}{r^2}\cos\theta.$$

The first of these results implies that the field inside the sphere is uniform of intensity $\boldsymbol{E} = 3\boldsymbol{E}_0/(\varepsilon + 2)$. The second result implies that the polarized sphere contributes to the external field like a dipole at O of moment $4\pi\varepsilon_0 a^3\boldsymbol{E}_0(\varepsilon - 1)/(\varepsilon + 2)$. □

Problem 2.7 A dipole of moment $\boldsymbol{m}$ is placed at the centre of a spherical cavity in an infinite dielectric of constant ε. Show that the field intensity in the dielectric is the same as that produced by a dipole of moment $3\boldsymbol{m}/(2\varepsilon + 1)$ in empty space.

Solution. Take spherical polar coordinates (r, θ) as indicated in Fig. 2.4. Inside the cavity, the potential will be the sum of a part due to the dipole and a part due to the polarized dielectric. We assume

$$V = V_1 = \frac{(m\cos\theta)}{(4\pi\varepsilon_0 r^2)} + Ar\cos\theta.$$

In the dielectric, the potential must vanish at infinity and we shall assume $V = V_2 = B\cos\theta/r^2$.

Across the wall of the cavity $r = a$, V and D_n are continuous. These conditions lead to the equations

$$\frac{m}{4\pi\varepsilon_0 a^2} + Aa = \frac{B}{a^2}, \qquad \frac{m}{2\pi\varepsilon_0 a^3} - A = \frac{2\varepsilon B}{a^3}.$$

Hence, $A = -m(\varepsilon - 1)/[4\pi\varepsilon_0 a^3(2\varepsilon + 1)]$, $B = 3m/[4\pi\varepsilon_0(2\varepsilon + 1)]$. The potential in the dielectric is therefore given by

$$V = \frac{3m\cos\theta}{4\pi\varepsilon_0(2\varepsilon + 1)r^2}$$

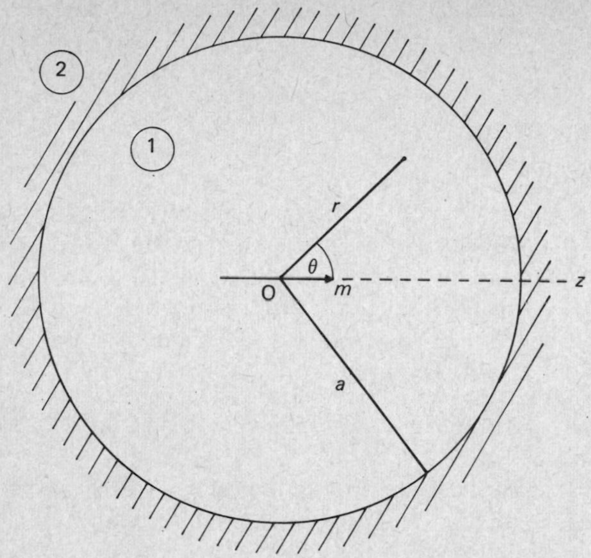

Fig. 2.4

and this is the potential of a dipole of moment $3\boldsymbol{m}/(2\varepsilon+1)$ placed at O in empty space. □

2.4 Images

Problem 2.8 The semi-infinite region on one side of a plane is occupied by dielectric of constant ε. A point charge q is placed outside the dielectric at a distance d from the plane. Find the force on q.

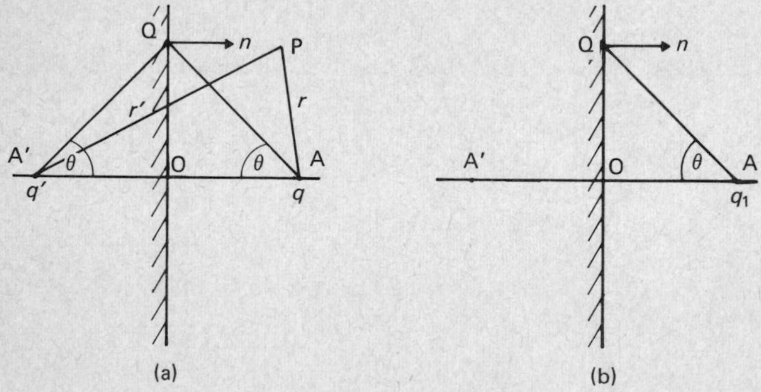

Fig. 2.5

Solution. If q is at A (Fig. 2.5), let A′ be the optical image of A in the plane face of the dielectric. We shall assume that the field potential outside the dielectric can be generated by point charges q at A and q' at A′. Both charges are to be thought of as situated in empty space. Thus, if r, r' are distances measured from A, A′ respectively, then

$$V = \frac{1}{4\pi\varepsilon_0}\left(\frac{q}{r}+\frac{q'}{r'}\right), \qquad r' \geqslant r. \tag{2.13}$$

Inside the dielectric, V will be supposed generated by an image charge q_1 at A (also regarded as being alone in space). Thus $V = q_1/(4\pi\varepsilon_0 r)$, $r' \leqslant r$.

V must be continuous at the plane face, and since $r = r'$ at Q,

$$q+q' = q_1. \tag{2.14}$$

Outside the dielectric at Q, the component of $\boldsymbol{E}$ in the direction of the normal $\boldsymbol{n}$ due to the charges q, q' is

$$\frac{q'}{4\pi\varepsilon_0 r'^2}\cos\theta - \frac{q}{4\pi\varepsilon_0 r^2}\cos\theta.$$

Multiplying by ε_0, this gives D_n. Inside the dielectric at Q, $E_n = -q_1\cos\theta/4\pi\varepsilon_0 r^2$. Hence, $D_n = -\varepsilon q_1\cos\theta/4\pi r^2$. But D_n is continuous at Q and, therefore,

$$q'-q = -\varepsilon q_1. \tag{2.15}$$

Solving equations (2.14), (2.15), we find that

$$q_1 = \frac{2}{\varepsilon+1}q, \qquad q' = -\frac{\varepsilon-1}{\varepsilon+1}q. \tag{2.16}$$

By appeal to a uniqueness theorem, it now follows that the fields in the two regions are correctly generated by these two systems of image charges.

In particular, if the dielectric is removed and the image q' is introduced, the field outside the dielectric will be unaltered. Hence, the force on q is equal to the attraction due to q', namely $(\varepsilon-1)q^2/[16\pi\varepsilon_0(\varepsilon+1)d^2]$. □

Problem 2.9 If, in the previous problem, a real point charge e is placed in the dielectric at the point A′, calculate the force on q.

Solution. We suppose the potential outside the dielectric to be generated by point charges q, q' at A, A′ respectively, as before. Inside the dielectric we suppose the potential generated by charges q_1, e/ε at A, A′ respectively; it is understood that all charges are being imagined situated in a vacuum, so that the potential arising from this pair of charges is

$$V = \frac{q_1}{4\pi\varepsilon_0 r}+\frac{e}{4\pi\varepsilon\varepsilon_0 r'}; \tag{2.17}$$

comparing the second term with equation (2.7), it will be seen that the behaviour of V in the neighbourhood of the charge e at A′ has been generated correctly.

We now require that the potential functions (2.13), (2.17) should be continuous across the plane $r = r'$ and, further, that the associated D_n should be continuous. These conditions lead to the equations

$$q+q' = q_1+e/\varepsilon, \qquad q'-q = e-\varepsilon q_1 .$$

Solving for q', q_1, we find that

$$q' = \frac{2}{\varepsilon+1}e-\frac{\varepsilon-1}{\varepsilon+1}q, \qquad q_1 = \frac{2}{\varepsilon+1}q+\frac{\varepsilon-1}{\varepsilon(\varepsilon+1)}e.$$

The force on q is the attraction of q', i.e.

$$\frac{qq'}{16\pi\varepsilon_0 d^2} = \frac{q[(\varepsilon-1)q-2e]}{16\pi\varepsilon_0 d^2(\varepsilon+1)}.$$

□

Problem 2.10 An infinite line charge q is placed parallel to the axis of an infinite circular cylindrical dielectric of radius a. If the distance between the line charge and the cylinder axis is $f(> a)$, calculate the force per unit length on the charge.

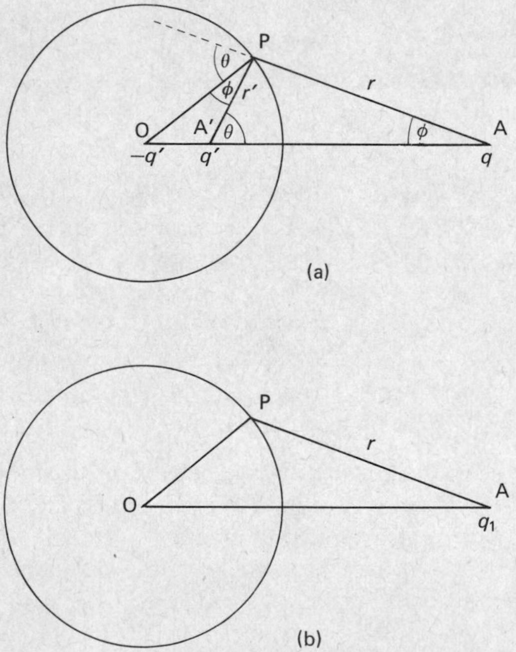

Fig. 2.6

Solution. It will be assumed that the potential outside the cylinder can be generated by the actual line charge q through the point A (Fig. 2.6(a)) and parallel image line charges (i) q' through the point A′ inverse to A in the cylinder and (ii) $-q'$ through O along the cylinder axis. For the actual field, the flux of $\boldsymbol{D}$ out of any length of the cylinder must be zero, since it contains no charge. The overall image charge within the cylinder must be taken to be zero, therefore, as we have assumed. The potential inside the cylinder will be supposed to be generated by a line charge q_1 through A (Fig. 2.6(b)). As usual, this charge is thought of as being alone in space.

The potential must be continuous at all points P on the cylinder. At such a point just outside the cylinder, the potential is being assumed to be $(-q\log r - q'\log r' + q'\log a + C)/2\pi\varepsilon_0$. The potential due to a line charge is always arbitrary to the extent of an added constant and advantage has been taken of this to introduce the constant C. The potential inside the cylinder at P is being assumed to be $-q_1\log r/2\pi\varepsilon_0$. The condition for continuity is accordingly $(q-q_1)\log r + q'\log r' = q'\log a + C$. But, by similar triangles, $r' = ar/f$ and, thus, this condition can be written $(q-q_1+q')\log r = q'\log f + C$. If this is to be satisfied for all r, it is necessary to take

$$q_1 \quad q' = q \tag{2.18}$$

and $C = -q'\log f$.

Continuity of D_n at P requires that

$$\frac{q}{2\pi r}\cos\theta - \frac{q'}{2\pi r'}\cos\phi + \frac{q'}{2\pi a} = \frac{\varepsilon q_1}{2\pi r}\cos\theta. \tag{2.19}$$

By projecting OA on to AP, it follows from the figure that $f\cos\phi = a\cos\theta + r$. Since $f = ar/r'$, this implies that

$$\frac{1}{r'}\cos\phi = \frac{1}{r}\cos\theta + \frac{1}{a}.$$

Equation (2.19) according reduces to

$$(q - q' - \varepsilon q_1)\frac{\cos\theta}{2\pi r} = 0.$$

This is satisfied at all points on the cylinder, provided

$$q' + \varepsilon q_1 = q. \tag{2.20}$$

Solving equations (2.18), (2.20), we find that

$$q' = -\frac{\varepsilon - 1}{\varepsilon + 1}q, \qquad q_1 = \frac{2}{\varepsilon+1}q.$$

The field intensity at A due to the image charges within the cylinder is now found to be

$$\frac{q}{2\pi\varepsilon_0}\cdot\frac{\varepsilon-1}{\varepsilon+1}\cdot\frac{a^2}{f(f^2-a^2)}$$

directed towards O. Thus, the force per unit length of the line charge at A is $a^2q^2(\varepsilon-1)/[2\pi\varepsilon_0(\varepsilon+1)f(f^2-a^2)]$. □

2.5 Magnetostatic Field The sources of the magnetostatic field are certain atoms which behave towards one another in the same manner as electric dipoles. However, the physical basis of this magnetic property is quite different from that of an electric dipole; it is caused by the circulation of electric charge within the atom. Nevertheless, the similarity of behaviour is so close that the mathematical theory we have developed to describe the phenomenon of electric polarization can be applied with only small modification to magnetically polarized materials.

Although isolated magnetic poles have never been observed, it is mathematically convenient to base the theory on a law of force between two poles of strengths p_1, p_2. This is the analogue of equation (1.3) and is written

$$\boldsymbol{F} = \frac{\mu_0 p_1 p_2}{4\pi r^2}\hat{\boldsymbol{r}}, \tag{2.21}$$

where $\mu_0 = 1{\cdot}257\times10^{-6}$ is called the *permeability of free space* and p_1, p_2 are measured in SI units (see p. 95).*

In free space, the force per unit pole is denoted by $\boldsymbol{B}$ and is termed the *magnetic induction.* Thus, for a hypothetical pole of strength p at the origin, it follows from equation (2.21) that $\boldsymbol{B} = \mu_0 p\hat{\boldsymbol{r}}/4\pi r^2$. Also, the couple which acts upon a magnetic dipole of moment $\boldsymbol{m}$ placed in a field $\boldsymbol{B}$ is $\boldsymbol{m}\times\boldsymbol{B}$.

In free space, the *magnetic intensity* $\boldsymbol{H}$ is defined to be $\boldsymbol{B}/\mu_0$. Thus, for a pole of strength p at the origin, $\boldsymbol{H} = p\hat{\boldsymbol{r}}/4\pi r^2$. As has been remarked in §1.2, this particular field is conservative and can be derived from the magnetic potential $\Omega = p/4\pi r$ by application of the equation

$$\boldsymbol{H} = -\operatorname{grad}\Omega. \tag{2.22}$$

Since any magnetostatic field can be generated by a distribution of poles (in dipole pairs), equation (2.22) is certainly valid at all points in free space. In particular, for an isolated dipole of moment $\boldsymbol{m}$ at the origin,

---

* μ_0 is placed in the numerator to permit $\boldsymbol{B}$ to be taken as the force per unit pole instead of $\boldsymbol{H}$. This is the accepted modern convention.

$$\Omega = \frac{m\cos\theta}{4\pi r^2} = \frac{\boldsymbol{m}\,.\,\boldsymbol{r}}{4\pi r^3}. \tag{2.23}$$

Comparing this equation with equation (1.9), it should be observed that the potential of an electrostatic distribution can always be converted into the potential of the corresponding distribution of magnetic poles by setting $\varepsilon_0 = 1$.

It can now be shown, exactly as in the corresponding electrostatic situation that, if $\boldsymbol{M}$ is the magnetic moment per unit volume within a magnetic material, the field generated externally is identical with that of a distribution of poles of volume density $-\operatorname{div}\boldsymbol{M}$ over the region occupied by the material and a distribution of poles of density $\boldsymbol{M}\,.\,\boldsymbol{n}$ over its surface ($\boldsymbol{n}$ is the unit outwards normal). This is the Poisson distribution for the material. $\boldsymbol{M}$ is termed the *magnetization.* Within the material, the potential Ω and intensity $\boldsymbol{H}$ are defined to be those of the Poisson distribution. This ensures that equation (2.22) is valid everywhere.

Since $\boldsymbol{H} = p\hat{\boldsymbol{r}}/4\pi r^2$ for a pole p at the origin, the flux of $\boldsymbol{H}$ out of any closed surface in a field due to a hypothetical distribution of poles is equal to the total magnetic charge within the surface. As shown in § 1.4, the differential form of this result is $\operatorname{div}\boldsymbol{H} = \rho$, where ρ is the density of magnetic charge. Application of this result to the Poisson distribution leads to the equation $\operatorname{div}\boldsymbol{H} = -\operatorname{div}\boldsymbol{M}$; hence,

$$\operatorname{div}\boldsymbol{B} = 0 \tag{2.24}$$

where

$$\boldsymbol{B} = \mu_0(\boldsymbol{H}+\boldsymbol{M}). \tag{2.25}$$

This definition of $\boldsymbol{B}$ within a magnetic material is consistent with the equation $\boldsymbol{B} = \mu_0\boldsymbol{H}$ in free space.

It can be proved that $\boldsymbol{H}$ is the intensity within a small needle-shaped cavity cut inside the material with its axis parallel to $\boldsymbol{M}$. Thus, the force on unit pole in such a cavity is $\mu_0\boldsymbol{H}$ (*not* $\boldsymbol{B}$). The intensity inside a small penny-shaped cavity whose axis is aligned with $\boldsymbol{M}$ is found to be $\boldsymbol{H}+\boldsymbol{M}$. Thus, the force on unit pole in the cavity is $\mu_0(\boldsymbol{H}+\boldsymbol{M}) = \boldsymbol{B}$.

Equation (2.24) is the differential form of Gauss's law for magnetostatic fields; this states that the flux of $\boldsymbol{B}$ out of any closed surface (which may intersect magnetized materials) always vanishes.

Problem 2.11 A sphere with centre at the origin and radius a is magnetized so that the components of magnetization are $\lambda(x, y, z+a)$. A dipole of moment m is placed at the point $(c, 0, 0)$ $(c > a)$ with its axis parallel to Ox. Calculate the couple acting upon it.

Solution. Since $\boldsymbol{M} = \lambda(x, y, z+a)$, $\operatorname{div} \boldsymbol{M} = 3\lambda$; thus the Poisson distribution over the interior of the sphere has density -3λ. If $\boldsymbol{r} = (x, y, z)$ is the position vector of a point, $\boldsymbol{M} = \lambda(\boldsymbol{r}+a\boldsymbol{k})$, where $\boldsymbol{k}$ is the unit vector along Oz. Thus, the radial component of $\boldsymbol{M}$ is $\boldsymbol{M}.\boldsymbol{r}/r = \lambda(r+a\cos\theta)$, where θ is the spherical polar angle between $\boldsymbol{r}$ and $\boldsymbol{k}$. Hence, on the sphere's surface, $M_n = \lambda a(1+\cos\theta)$ and this defines the surface Poisson distribution.

The volume charge of density -3λ and the part of the surface charge of density λa can, for the purpose of calculating the external field, be condensed to a point charge at O of magnitude $\frac{4}{3}\pi a^3(-3\lambda)+4\pi a^2(\lambda a) = 0$. It follows that the external field is that generated by a surface charge of density $\lambda a \cos\theta$.

The potential of a spherical surface distribution of this type has already been found in Problem 2.1; setting $P = \lambda a$ and $\varepsilon_0 = 1$ in the result of this calculation, we get $\Omega = \lambda a^4 \cos\theta/3r^2$ as the external potential. Thus, the external field is the same as that due to a dipole of moment $4\pi\lambda a^4/3$ at O with its axis along Oz.

The radial and transverse components of $\boldsymbol{H}$ due to this dipole can now be found by calculating $-\partial\Omega/\partial r$ and $-\partial\Omega/r\,\partial\theta$. In particular, the intensity at $(c, 0, 0)$ is found to be $\lambda a^4/3c^3$ directed along zO. Thus, $\boldsymbol{B} = -\boldsymbol{k}\lambda\mu_0 a^4/3c^3$ at this point and the given dipole experiences a couple $\boldsymbol{m}\times\boldsymbol{B} = m\boldsymbol{i}\times\boldsymbol{B} = \boldsymbol{j}m\lambda\mu_0 a^4/3c^3$ ($\boldsymbol{i}$ and $\boldsymbol{j}$ are unit vectors along the x- and y-axes. □

2.6 Harmonic Functions For most materials, $\boldsymbol{M}$ is directly proportional to $\boldsymbol{H}$ and we can write

$$\boldsymbol{M} = \chi\boldsymbol{H}, \tag{2.26}$$

where χ is the *magnetic susceptibility* (c.f. equation (2.1)). Using equation (2.25), this leads to the result

$$\boldsymbol{B} = \mu_0(1+\chi)\boldsymbol{H} = \mu\mu_0\boldsymbol{H}, \tag{2.27}$$

where $\mu = 1+\chi$ is termed the *permeability* of the material.

In such materials, the fundamental equations of the magnetic field are

$$\boldsymbol{H} = -\operatorname{grad}\Omega, \qquad \operatorname{div}\boldsymbol{B} = 0, \qquad \boldsymbol{B} = \mu\mu_0\boldsymbol{H}. \tag{2.28}$$

Ω therefore satisfies the equation

$$\operatorname{div}(\mu\operatorname{grad}\Omega) = 0 \tag{2.29}$$

and, if μ is constant, Ω satisfies Laplace's equation.

In ferromangetic materials, equation (2.26) is not valid and it is then usual to treat $\boldsymbol{M}$ as a given distribution. In such materials, the fundamental equations are

$$\boldsymbol{H} = -\operatorname{grad}\Omega, \qquad \operatorname{div}\boldsymbol{B} = 0, \qquad \boldsymbol{B} = \mu_0(\boldsymbol{H}+\boldsymbol{M}). \tag{2.30}$$

Ω then satisfies the Poisson equation.

$$\nabla^2\Omega = \operatorname{div}\boldsymbol{M}. \tag{2.31}$$

The boundary conditions at a surface of discontinuity separating two magnetic media are (i) Ω is continuous and (ii) B_n is continuous. The second of the conditions follows from Gauss's law by an argument analogous to that which led to equation (1.20).

Problem 2.12 A long, circular, cylindrical, soft iron pipe has internal radius a and external radius b. Its permeability is μ. It is placed with its axis perpendicular to a uniform field of intensity $\boldsymbol{H}_0$. Calculate the resultant field.

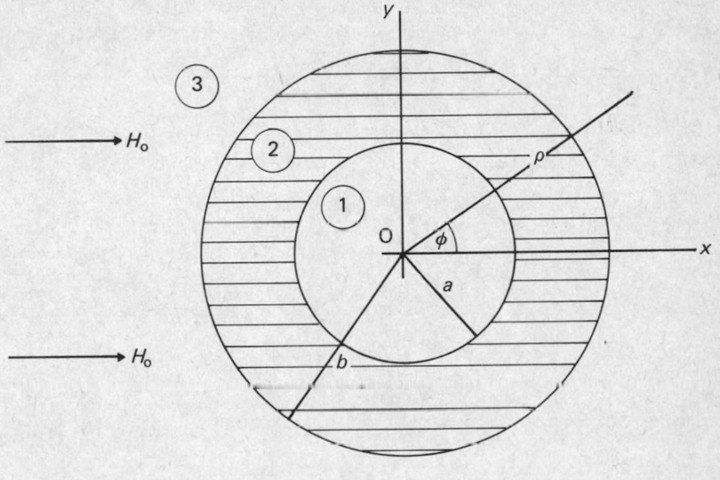

Fig. 2.7

Solution. Taking the axis of the pipe as z-axis and x-axis in the direction of the applied field, we introduce cylindrical polar coordinates (ρ, ϕ) as shown in Fig. 2.7. The potential of the uniform field is then $-H_0x = -H_0\,\rho\cos\phi$. We assume that this potential is modified by the induced magnetism in the pipe so that the potential in region 3 is given by $\Omega = \Omega_3 = -H_0\,\rho\cos\phi + A\cos\phi/\rho$. The added term vanishes at infinity and satisfies Laplace's equation as required. Inside the soft iron, we assume $\Omega = \Omega_2 = (B\rho + C/\rho)\cos\phi$, and, inside the hollow, we take $\Omega = \Omega_1 = D\rho\cos\phi$. Both Ω_1 and Ω_2 satisfy Laplace's equation and Ω_1 is finite at $\rho = 0$. Ω is continuous across $\rho = a$ and $\rho = b$ provided

$$-H_0 b + \frac{A}{b} = Bb + \frac{C}{b}, \qquad Ba + \frac{C}{a} = Da.$$

$B_n = \mu\mu_0 H_n = -\mu\mu_0 \,\partial\Omega/\partial\rho$ is continuous across these two surfaces provided

$$H_0 + \frac{A}{b^2} = \mu\left(\frac{C}{b^2} - B\right), \qquad \mu\left(\frac{C}{a^2} - B\right) = -D.$$

A, B, C, D can now be found. In particular

$$D = -\frac{4\mu H_0}{(\mu+1)^2 - (\mu-1)^2 a^2/b^2}.$$

It follows that the field in the hollow interior of the pipe is uniform and of intensity $4\mu \boldsymbol{H}_0/[\mu+1)^2 - (\mu-1)^2 a^2/b^2]$. For large values of μ, only the μ^2 terms in the denominator need be retained, and the intensity approximates to $4\boldsymbol{H}_0/[\mu(1-a^2/b^2)]$. For large μ, this will be small, showing that the magnetic material has a shielding effect. □

Problem 2.13 A long circular steel cylinder of radius a is magnetized uniformly in a direction perpendicular to its axis. Calculate the field.

Solution. Introduce cylindrical polar coordinates (ρ, ϕ) as indicated in Fig. 2.8, the x-axis being parallel to $\boldsymbol{M}$. Since $\boldsymbol{M}$ is constant, equation (2.31) reduces to Laplace's equation and Ω is harmonic everywhere.

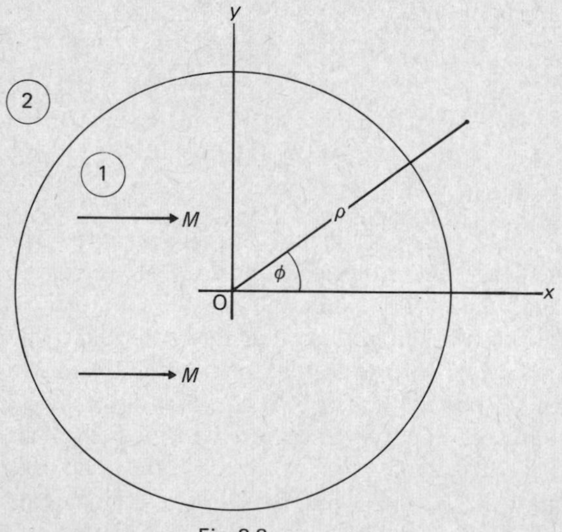

Fig. 2.8

Inside the cylinder, we take $\Omega = \Omega_1 = P\rho\cos\phi$; outside the cylinder, we assume $\Omega = \Omega_2 = Q\cos\phi/\rho$. Ω is continuous at the surface of the cylinder if $Pa = Q/a$. Inside the cylinder, $\boldsymbol{B}$ is given by equation (2.25) and its radial component is accordingly $\mu_0(M\cos\phi - \partial\Omega_1/\partial\rho) = \mu_0(M-P)\cos\phi$. Outside the cylinder, $\boldsymbol{B} = \mu_0\boldsymbol{H}$ and the radial component is $-\mu_0\,\partial\Omega_2/\partial\rho = \mu_0 Q\cos\phi/\rho^2$. Hence, B_n is continuous across $\rho = a$ provided $\mu_0(M-P) = \mu_0 Q/a^2$. Solving for P and Q, we get $P = \frac{1}{2}M$ and $Q = \frac{1}{2}Ma^2$.

It now follows that the external field is that of a magnetic doublet of moment $\pi a^2 M$ (the potential of an electric doublet of moment m is $m\cos\phi/2\pi\varepsilon_0\rho$ (Exercise 1, Chapter 1); for a magnetic doublet, put $\varepsilon_0 = 1$).

The internal field is uniform and of intensity $-\frac{1}{2}\boldsymbol{M}$. Note that this opposes the magnetization; it is termed the *demagnetizing field* and tends to destroy the magnetism. □

2.7 Forces between Dipoles If a magnetic dipole of moment $\boldsymbol{m}$ is placed in a field of induction $\boldsymbol{B}$, its potential energy is $-\boldsymbol{m}\,.\,\boldsymbol{B}$.

Problem 2.14 Calculate the forces of interaction between two dipoles whose axes lie on the same plane.

Solution. Let $\boldsymbol{m}$, $\boldsymbol{m}'$ be the moments of the two dipoles and let these be placed in relation to one another as indicated in Fig. 2.9. The forces acting upon $\boldsymbol{m}'$ are equivalent to a force acting at its centre $\boldsymbol{P}$ and a couple. Let G be the couple and let (R, T) be the components of the force along the line joining the dipoles and perpendicular to this line.

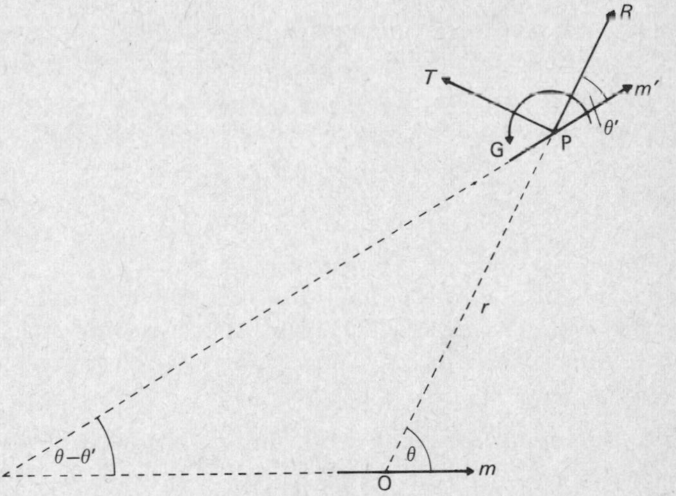

Fig. 2.9

Referring to equation (1.12), the field generated by $\boldsymbol{m}$ at $\boldsymbol{m}'$ is given by

$$\boldsymbol{B} = \frac{\mu_0}{4\pi}\left(\frac{3\boldsymbol{m}\,.\,\boldsymbol{r}}{r^5}\boldsymbol{r} - \frac{\boldsymbol{m}}{r^3}\right).$$

Thus, the potential energy of $\boldsymbol{m}'$ in this field is W, where

$$W = -\boldsymbol{m}'\,.\,\boldsymbol{B} = \frac{\mu_0}{4\pi}\left(\frac{1}{r^3}\boldsymbol{m}\,.\,\boldsymbol{m}' - \frac{3}{r^5}\boldsymbol{m}\,.\,\boldsymbol{r}\,\boldsymbol{m}'\,.\,\boldsymbol{r}\right),$$

$$= \frac{\mu_0 mm'}{4\pi r^3}[\cos(\theta-\theta') - 3\cos\theta\cos\theta'],$$

$$= \frac{\mu_0\,mm'}{4\pi r^3}(\sin\theta\sin\theta' - 2\cos\theta\cos\theta'). \tag{2.32}$$

Now suppose that $\boldsymbol{m}'$ is displaced slightly so that θ increases by $d\theta$, θ' increases by $d\theta'$ and r increases by dr. Then P is displaced by dr in the direction OP and by $r\,d\theta$ in the perpendicular direction; $\boldsymbol{m}'$ also rotates through an angle $d\theta - d\theta'$. Thus, the work done by the forces of the system is $R\,dr + Tr\,d\theta + G(d\theta - d\theta')$. This will equal the decrease in the potential energy $-dW$. But

$$dW = \frac{\partial W}{\partial \theta}d\theta + \frac{\partial W}{\partial \theta'}d\theta' + \frac{\partial W}{\partial r}dr.$$

Hence $\quad Tr + G = -\dfrac{\partial W}{\partial \theta}, \quad G = \dfrac{\partial W}{\partial \theta'}, \quad R = -\dfrac{\partial W}{\partial r}.$

Substitution for W from equation (2.32), now yields the results

$$\left.\begin{aligned} G &= \frac{\mu_0\,mm'}{4\pi r^3}(\sin\theta\cos\theta' + 2\cos\theta\sin\theta') \\ R &= \frac{3\mu_0\,mm'}{4\pi r^4}(\sin\theta\sin\theta' - 2\cos\theta\cos\theta') \\ T &= -\frac{3\mu_0\,mm'}{4\pi r^4}\sin(\theta+\theta'). \end{aligned}\right\} \tag{2.33}$$

□

Problem 2.15 Two dipoles, of moments m and $6m$, are pivoted at points P and Q and are free to rotate about their centres in a plane containing PQ. A uniform field $\boldsymbol{B}$ is parallel to this plane and perpendicular to PQ. Show that the position of equilibrium in which the dipoles are aligned with $\boldsymbol{B}$ is stable if $4\pi r^3 B > 9\mu_0 m$, where r = PQ.

Solution. Suppose the dipole m is first introduced into the field $\boldsymbol{B}$; the potential energy generated is $-mB\cos\theta$ (Fig. 2.10). The dipole $6m$ is then introduced into the combined field of m and $\boldsymbol{B}$. Its energy in the field $\boldsymbol{B}$ is

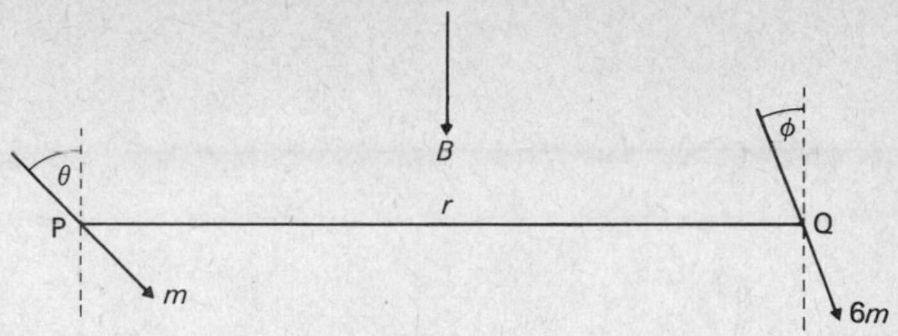

Fig. 2.10

$-6mB\cos\phi$ and its energy in the field of m follows from equation (2.32) by putting $m' = 6m$, $\theta' = \frac{1}{2}\pi - \phi$, and replacing θ by $\frac{1}{2}\pi - \theta$. Hence, the total potential energy of the system is

$$W = -mB\cos\theta - 6mB\cos\phi + \frac{3\mu_0 m^2}{2\pi r^3}(\cos\theta\cos\phi - 2\sin\theta\sin\phi).$$

It is now easy to verify that $\partial W/\partial\theta = \partial W/\partial\phi = 0$ when $\theta = \phi = 0$ and hence that W is stationary in this configuration. Thus the dipoles are in equilibrium when they are aligned with the applied field. Also,

$$\frac{\partial^2 W}{\partial\theta^2} = mB\cos\theta + \frac{3\mu_0 m^2}{2\pi r^3}(2\sin\theta\sin\phi - \cos\theta\cos\phi),$$

$$\frac{\partial^2 W}{\partial\phi^2} = 6mB\cos\phi + \frac{3\mu_0 m^2}{2\pi r^3}(2\sin\theta\sin\phi - \cos\theta\cos\phi),$$

$$\frac{\partial^2 W}{\partial\theta\partial\phi} = \frac{3\mu_0 m^2}{2\pi r^3}(\sin\theta\sin\phi - 2\cos\theta\cos\phi).$$

Thus, when $\theta = \phi = 0$,

$$\frac{\partial^2 W}{\partial\theta^2} = mB(1-3k), \qquad \frac{\partial^2 W}{\partial\phi^2} = 3mB(2-k),$$

$$\frac{\partial^2 W}{\partial\theta^2}\frac{\partial^2 W}{\partial\phi^2} - \left(\frac{\partial^2 W}{\partial\theta\partial\phi}\right)^2 = 3m^2B^2(1+k)(2-9k),$$

where $k = \mu_0 m/2\pi r^3 B$. For stable equilibrium, W must be a minimum. The condition for this is that all the expressions just calculated must be positive, i.e. $-1 < k < \frac{2}{9}$. Since k is positive, the condition required is $4\pi r^3 B > 9\mu_0 m$. □

Problem 2.16 The region $x < 0$ is occupied by material of permeability μ. A dipole of moment m is placed at the point A$(a, 0, 0)$ $(a > 0)$ with its axis perpendicular to the plane face of the material. Calculate the force on the dipole.

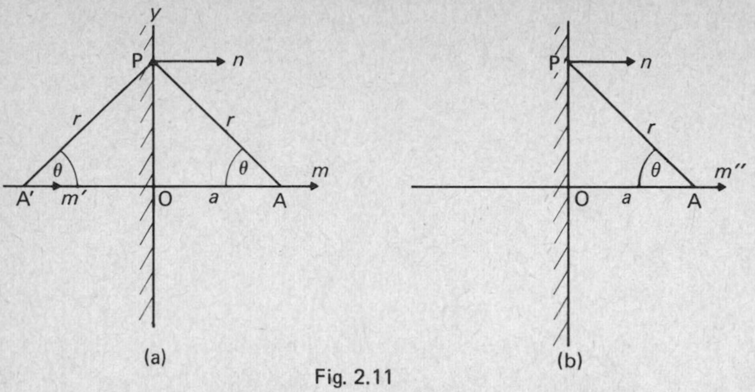

Fig. 2.11

Solution. The potential in the region $x > 0$ is assumed generated by the actual dipole at A and an image dipole m' at A′, the optical image of A in the plane face of the material (Fig. 2.11a). The potential in the region $x < 0$ is assumed generated by an image dipole m'' at A, regarded as situated in empty space (Fig. 2.11b).

The potential is continuous across the face at a point P, provided

$$\frac{m}{4\pi r^2}\cos(\pi-\theta)+\frac{m'}{4\pi r^2}\cos\theta = \frac{m''}{4\pi r^2}\cos(\pi-\theta),$$

i.e.

$$-m+m' = -m''. \tag{2.34}$$

The component in the direction of the normal $\boldsymbol{n}$ of the intensity at P due to m'' is calculated from equation (1.12) (with $\varepsilon_0 = 1$) to be

$$H_n = \frac{m''}{4\pi r^3}(3\cos^2\theta-1).$$

Since $\boldsymbol{B} = \mu\mu_0\boldsymbol{H}$ inside the material at P,

$$B_n = \frac{\mu\mu_0 m''}{4\pi r^3}(3\cos^2\theta-1). \tag{2.35}$$

Similarly, B_n generated by the dipoles m, m' at P is given by

$$B_n = \frac{\mu_0}{4\pi r^3}(m+m')(3\cos^2\theta-1). \tag{2.36}$$

Equations (2.35), (2.36), show that B_n is continuous provided

$$m+m' = \mu m''. \tag{2.37}$$

Solving equations (2.34), (2.37), we find that

$$m' = \frac{\mu-1}{\mu+1}m, \qquad m'' = \frac{2}{\mu+1}m.$$

Equations (2.33) now show that m is attracted towards m' by a force

$$\frac{3\mu_0 mm'}{32\pi a^4} = \frac{3\mu_0 m^2}{32\pi a^4}\cdot\frac{\mu-1}{\mu+1}. \qquad \square$$

EXERCISES

1. A spherical shell having internal radius a and external radius b is formed from insulating material whose dielectric constant is ε. A point charge is placed at the centre. Show that the external field is the same as the field which would be present if the shell were removed. The charge is now replaced by a dipole of moment m. Show that the external field is the same as if the shell were absent and the dipole moment were λm, where

$$\frac{1}{\lambda} = 1+\frac{2(\varepsilon-1)^2}{9\varepsilon}\left(1-\frac{a^3}{b^3}\right).$$

2. The planes $x = 0$, $x = d$ are occupied by conductors and the region between them is occupied by an insulator whose dielectric constant is given by $\varepsilon = (1-x/2d)^{-1}$. The conductor $x = 0$ is earthed and the other is raised to a potential ϕ. Show that the capacitance per unit area of the plate $x = d$ is $4\varepsilon_0/3d$.

3. An infinite line charge q lies along the axis of an infinite circular cylindrical dielectric of radius a. A parallel infinite line charge $-q$ is placed in the dielectric a distance f from its axis. Prove that the force per unit length which acts upon the first charge has magnitude

$$\frac{q^2}{2\pi\varepsilon\varepsilon_0}\left[\frac{1}{f}+\frac{\varepsilon-1}{\varepsilon+1}\frac{f}{a^2}\right].$$

Calculate the force per unit length on the other charge and explain why it is not equal and opposite to the force on the axial charge.

4. A sphere of radius a is magnetized so that the intensity of magnetization at the point (x, y, z) is in the direction of the z-axis and of magnitude αz, where α is a constant and the origin is at the centre of the sphere. Calculate the Poisson distribution and show that the magnetic potential outside the sphere is given by $\Omega = 2\alpha a^5 P_2(\cos\theta)/15r^3$.

5. Employing spherical polar coordinates (r, θ, ϕ), a medium has permeability $\mu_1(a/r)^2$, where μ_1 and a are constants. Show that solutions for the magnetic potential in the medium of the form $\Omega = F(r)\cos\theta$ exist and that $F(r) = Ar^2+B/r$. A sphere of radius a formed from this material, whose centre is the pole $r = 0$, is placed in a uniform field of intensity H.

Show that the potential inside the sphere is given by

$$\Omega = -\frac{3Hr^2 \cos\theta}{2(\mu_1+1)a}.$$

6. Three equal magnetic dipoles are pivoted at the vertices of an equilateral triangle, and can rotate in the plane of the triangle. They are constrained by a frictionless mechanism in such a way that their axes make equal angles, all in the same sense of rotation, with the inward bisectors of the angles at which they are situated. Show that there are four positions of equilibrium and that the two symmetrical positions are unstable.

Chapter 3

Steady Current Flows

3.1 Kirchhoff's Laws In this section, we consider steady electric currents in networks whose *branches* are wires meeting at points called the *nodes*. The strength of the current in a wire is measured by the rate at which charge flows across any cross-section of the wire. If n wires carrying currents $i_1, i_2, \ldots, i_n$ meet at a node, then since, under steady conditions, there is no accumulation of charge at the point,

$$i_1 + i_2 + \ldots + i_n = 0; \tag{3.1}$$

it is here assumed that currents directed towards the node are given a positive sign and currents directed away are given a negative sign. This is *Kirchhoff's First Law*.

Since there must be a positive component of electric field in the direction of the current to cause the charge to flow, the potential will decrease along a wire in the direction of the current. According to *Ohm's Law*, for any given wire, the potential drop along its length is proportional to the current flowing; thus, if i is the current and V is the potential drop, then $V = Ri$, where R is a constant for the wire called its *resistance*.

Consider a closed circuit formed by branches of a network AB, BC, CD, ..., KA. The electric potential will vary around the circuit, but must eventually return to its initial value upon return to the point A. It follows that the sum of the drops in potential along the wires of the circuit must be balanced by the potential rises across the terminals of the various batteries or other sources of current which are present in the circuit. It is found experimentally that the rise in potential across such a source of current is given by an expression $E - ri$, where E and r are constants for the source called the *electromotive force* (e.m.f.) and the *internal resistance* respectively, and i is the current in the sense of the potential rise. Thus, for any closed circuit in the network, it is necessary that

$$\Sigma\, Ri = \Sigma\,(E - ri), \tag{3.2}$$

or

$$\Sigma\, Ri = \Sigma\, E, \tag{3.3}$$

If it is understood that the set of resistances R includes the internal resistances of the sources. This it *Kirchhoff's Second Law*.

Problem 3.1 Eight equal wires, each of resistance r, form a pyramid ABCDE whose vertex is A and whose base is the square BCDE. Show that, if current is introduced at A and removed at P, a point on BC such that BP $= \lambda$BC, the resistance of the network between A and P is $7r(1+\lambda-\lambda^2)/15$.

Solution. Denote the currents in AB, AC, AD, AE, DE by v, w, x, y, z respectively (Fig. 3.1). The currents in EB, DC, BP, CP, can then be found (see figure) by application of the first law. Applying the second law to the

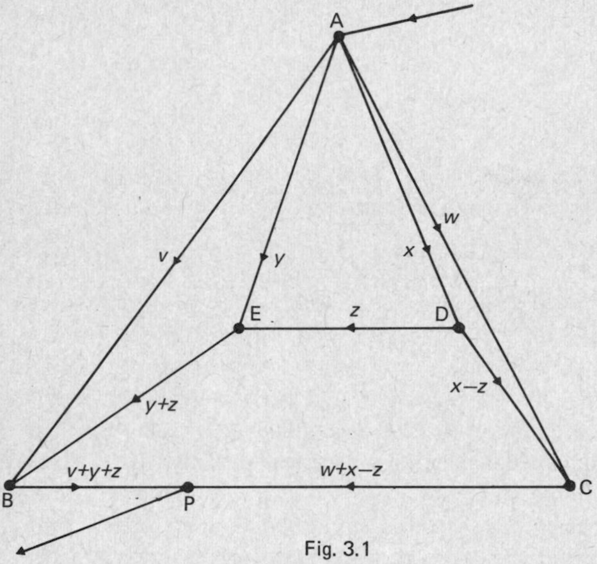

Fig. 3.1

circuits AEB, ADE, ADC, ABC (taking the resistances of BP and CP to be λr and $(1-\lambda)r$ respectively), the following equations are derived:

$$2y+z-v = 0; \qquad x+z-y = 0;$$

$$2x-z-w = 0; \qquad (\lambda+1)v+\lambda y+z-(2-\lambda)w-(1-\lambda)x = 0.$$

Solving for v, w, x, y in terms of z, we obtain

$$(1-2\lambda)x = (2+\lambda)z, \qquad (1-2\lambda)y = (3-\lambda)z,$$

$$(1-2\lambda)w = (3+4\lambda)z, \qquad (1-2\lambda)v = (7-4\lambda)z.$$

The current entering at A and leaving at P is $v+w+x+y = 15z/(1-2\lambda)$. The potential drop along ABP is $vr+(v+y+z)\lambda r = 7(1+\lambda-\lambda^2)zr/(1-2\lambda)$. The result stated now follows by dividing the potential difference between A and P by the total current entering the network. □

Problem 3.2 In the ladder network shown in Fig. 3.2, the sides of every square have resistance R and the ladder extends indefinitely to the right.

Current i enters at A and leaves at B. Find the equivalent resistance of the network between A and B.

Solution. If PQRS is the rth square counting from the end AB of the ladder, let i_r be the current in the branches PQ and RS. Then, applying the first law at the nodes, P, Q, we see that the currents in PS and QR are

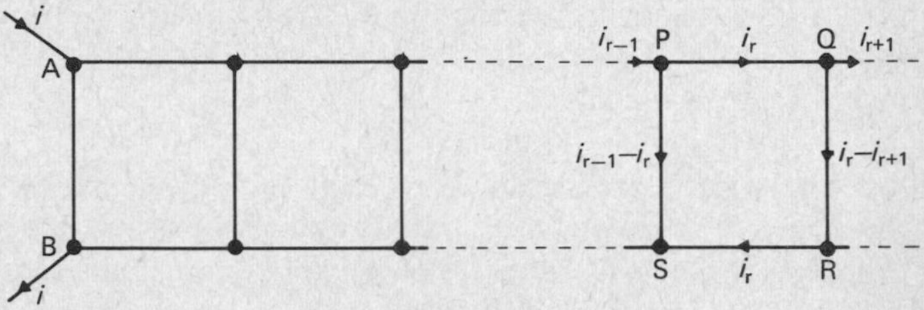

Fig. 3.2

$i_{r-1}-i_r$ and i_r-i_{r+1} respectively. Applying the second law to the circuit PQRSP, we obtain

$$R(i_r+i_r-i_{r+1}+i_r-i_{r-1}+i_r) = 0,$$

or

$$i_{r+1}-4i_r + i_{r-1} = 0.$$

The characteristic equation for this linear recurrence relationship is $\lambda^2-4\lambda+1 = 0$ and the characteristic roots are accordingly $2\pm\sqrt{3}$. Its general solution is therefore

$$i_r = A(2+\sqrt{3})^r+B(2-\sqrt{3})^r.$$

It is clear that i_r must remain bounded as $r \to \infty$ and this will only be so if $A = 0$. Further, $i_0 = i$ and, hence, $B = i$. Thus $i_r = i(2-\sqrt{3})^r$.

Since $i_1 = i(2-\sqrt{3})$, it follows that the current in AB is $i-i_1 = i(\sqrt{3}-1)$. Thus, the potential drop from A to B is $iR(\sqrt{3}-1)$. This implies that the whole network behaves like a single resistor of resistance $R(\sqrt{3}-1)$. This is called the *equivalent resistance* of the network. □

3.2 Heat Generation in a Network If a steady current i is flowing in a wire and V is the potential drop along the wire, in time t a charge it is transferred from the high potential end of the wire to the low potential end and so loses potential energy itV. This loss is compensated by a corresponding increase in the kinetic energy of the charge, which is then transferred by impact to the molecules of the wire and appears as heat. Thus, the rate of generation of heat energy is $iV = i^2R$, if R is the wire's resistance.

Problem 3.3 A source of current has electromotive force E and internal resistance r. Show that the maximum rate at which heat can be generated in a resistor connected to this supply is $E^2/4r$.

Solution. If a resistance R is connected to the supply, the current i which flows is given by the second law $E = (R+r)i$. Thus, the rate of generation of heat in R is $i^2R = E^2R/(R+r)^2$. Differentiating this expression with respect to R and setting the result to zero, it is found that $R = r$ makes the rate of heat generation a maximum. With this value of R, the rate of heat generation is $E^2/4r$, as stated. □

3.3 Steady Flows in Conducting Solids The pattern of charge flow within a solid conductor is specified by giving the *current density* vector $\boldsymbol{j}$ at all points. This vector is in the direction of flow and its magnitude is equal to di/dS, where di is the current flowing across a small surface element dS which is perpendicular to the flow. The rate of flow of charge across a surface S is $\int_S \boldsymbol{j}\,.\,\boldsymbol{n}\,dS$, where $\boldsymbol{n}$ is the unit normal to S in the sense in which the flow is to be calculated. If S is closed and $\boldsymbol{n}$ is outwards from the region Γ enclosed by S, this rate of flow must equal the rate at which the total charge inside S is decreasing (assuming there are no electrodes inside S). Thus, if ρ is the charge density

$$\int_S \boldsymbol{j}\,.\,\boldsymbol{n}\,dS = -\frac{d}{dt}\int_\Gamma \rho\,dv.$$

Applying the divergence theorem to the surface integral and differentiating under the integral sign in the volume integral, this gives

$$\int_\Gamma \left(\frac{\partial\rho}{\partial t}+\operatorname{div}\boldsymbol{j}\right)dv = 0.$$

Since Γ is arbitrary, this implies

$$\frac{\partial\rho}{\partial t}+\operatorname{div}\boldsymbol{j} = 0, \tag{3.4}$$

which is the *equation of continuity*.

For a steady flow $\partial\rho/\partial t = 0$, and hence

$$\operatorname{div}\boldsymbol{j} = 0. \tag{3.5}$$

Also, for a steady flow, Ohm's law in the differential form

$$\boldsymbol{j} = \sigma\boldsymbol{E} \tag{3.6}$$

is valid; σ is called the *conductivity*. Finally, for a steady flow it is assumed that $\boldsymbol{E}$ is derivable from a potential V by the equation

$$\boldsymbol{E} = -\operatorname{grad} V. \tag{3.7}$$

Regions from which current is supplied to a conductor, or into which current flows from a conductor, are called *electrodes*. These are usually assumed to have infinite conductivity and, hence, by equation (3.6), $\boldsymbol{E}$ will vanish within an electrode; equation (3.7) then implies that V is constant over an electrode. The net current supplied or received by an electrode is the flux of $\boldsymbol{j}$ across its surface and is called the *strength* of the electrode.

Problem 3.4 A single spherical electrode of strength i is set into an infinite conductor of conductivity σ. Calculate $\boldsymbol{j}$, $\boldsymbol{E}$ and V.

Solution. The flow will be radial everywhere and the current across a concentric sphere of radius r will be the flux of $\boldsymbol{j}$ across this surface, namely $4\pi r^2 j$. Assuming steady state conditions, this must equal i, and hence $j = i/4\pi r^2$. This leads to the equation

$$\boldsymbol{j} = \frac{i}{4\pi r^3}\boldsymbol{r}. \tag{3.8}$$

Equation (3.6) now yields

$$\boldsymbol{E} = \frac{i}{4\pi\sigma r^3}\boldsymbol{r}, \tag{3.9}$$

and equation (3.7) then implies that

$$V = \frac{i}{4\pi\sigma r}. \tag{3.10}$$

Note that equations (3.8)–(3.10) have the same forms as equations (2.5)–(2.7) for a point charge in an infinite dielectric. □

Problem 3.5 A circular electrode of strength i is set in an infinite conducting plate of conductivity σ and thickness t. Calculate the equations of the flow.

Solution. The flow will be radially outwards from the electrode and the flux of $\boldsymbol{j}$ across a cylinder, coaxial with the electrode and of radius ρ, will be $2\pi\rho t j$. This must equal i and hence $j = i/(2\pi t\rho)$. The equations for $\boldsymbol{E}$ and V are now derived as before. Thus

$$\boldsymbol{j} = \frac{i}{2\pi t\rho}\hat{\boldsymbol{u}}, \qquad \boldsymbol{E} = \frac{i}{2\pi\sigma t\rho}\hat{\boldsymbol{u}}, \qquad V = -\frac{i}{2\pi\sigma t}\log\rho, \tag{3.11}$$

where $\hat{\boldsymbol{u}}$ is the unit radial vector. □

A similar problem is that of the infinite circular cylindrical electrode of strength i per unit length, set in an infinite solid conductor. The equations can be derived from (3.11) by setting $t = 1$.

Problem 3.6 Three perfectly conducting circular electrodes, each of small radius δ, are set into an infinite conducting plate at the vertices of an equilateral triangle of side a. If a current i enters at one electrode and currents $\frac{1}{2}i$ leave by the other two, show that the equivalent resistance between the electrodes is

$$\frac{3}{4\pi\sigma t}\log\left(\frac{a}{\delta}\right).$$

Solution. Since the electrodes are small, the potential at any point on the plate can be found by summation of contributions due to each (c.f. point

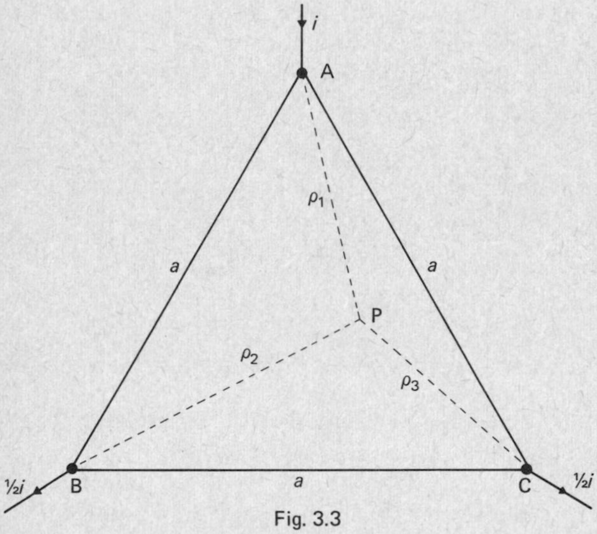

Fig. 3.3

charges). Thus, if P is a point on the plate distant ρ_1, ρ_2, ρ_3 from the three electrodes (Fig. 3.3), the potential at P will be given by

$$V = -\frac{i}{2\pi\sigma t}(\log\rho_1 - \tfrac{1}{2}\log\rho_2 - \tfrac{1}{2}\log\rho_3) = \frac{i}{4\pi\sigma t}\log\left(\frac{\rho_2\rho_3}{\rho_1^2}\right).$$

At the electrode A, $\rho_1 = \delta$, $\rho_2 = \rho_3 = a$, and hence

$$V_A = \frac{i}{4\pi\sigma t}\log\left(\frac{a^2}{\delta^2}\right).$$

At the electrode B, $\rho_1 = \rho_3 = a$, $\rho_2 = \delta$ and, therefore,

$$V_B = \frac{i}{4\pi\sigma t}\log\left(\frac{\delta}{a}\right).$$

Thus
$$(V_A - V_B)/i = \frac{3}{4\pi\sigma t}\log\left(\frac{a}{\delta}\right).$$

This is the equivalent resistance. □

3.4 Harmonic Functions and Images Any steady current flow is governed by equations (3.5)–(3.7). Comparison of these equations with the earlier electrostatic field equations (2.2)–(2.4) reveals immediately the parallelism which must exist between the two theories; for, by replacing $\boldsymbol{D}$ by $\boldsymbol{j}$, $\varepsilon\varepsilon_0$ by σ and setting $\rho = 0$, the electrostatic equations are transformed into those for a current flow. However, the boundary conditions for flow problems take a distinctive form and these will now be stated: (i) over the surface of an electrode, V is constant and, either V takes a given value (electrode's potential known), or the flux of $\boldsymbol{j}$ out of the surface takes a given value (electrode's strength known); (ii) over an insulated boundary, $j_n = 0$, since there is no flow from the conductor into the insulator; also, at a surface separating two conductors having different conductivities σ_1 and σ_2, V and j_n are usually assumed to be continuous; j_n *must* be continuous if there is to be no accumulation of charge at the interface, but the continuity of V depends upon there being no chemical interaction between the two conductors, which might lead to a *contact potential difference* being generated.

Eliminating $\boldsymbol{j}$ and $\boldsymbol{E}$ between equations (3.5)–(3.7), it is found that V satisfies the equation
$$\operatorname{div}(\sigma \operatorname{grad} V) = 0. \tag{3.12}$$

Thus, over any region for which σ is constant, V satisfies Laplace's equation and the methods of harmonic functions and images previously described become applicable.

Problem 3.7 A conducting sphere of radius a and conductivity σ_1 is embedded in an infinite conducting medium of conductivity σ_2, and modifies a steady current flow which would otherwise have a uniform current density $\boldsymbol{j}_0$. Show that the current density in the sphere is $3\sigma_1 \boldsymbol{j}_0/(\sigma_1 + 2\sigma_2)$ and calculate the density of charge on the spherical interface.

Solution. This problem is mathematically similar to the electrostatic problem of a dielectric sphere placed in a uniform field (Problem 2.6). Taking the pole of spherical polar coordinates at the centre of the sphere and the z-axis in the direction of $\boldsymbol{j}_0$, the potential associated with the uniform flow will be $V = -j_0 r\cos\theta/\sigma_2$ (for this corresponds to a uniform field of intensity $\boldsymbol{E} = j_0/\sigma_2$ and, hence, $\boldsymbol{j} = \sigma_2 \boldsymbol{E} = j_0$). We now assume that the potentials inside and outside the sphere are given by

$$V_1 = Br\cos\theta, \qquad V_2 = -\frac{j_0}{\sigma_2}r\cos\theta + \frac{A}{r^2}\cos\theta$$

respectively. The constants A, B are now chosen to ensure that $j_n = -\sigma(\partial V/\partial r)$ and V are continuous across the surface $r = a$; as in Problem 2.6, we find that

$$A = \frac{\sigma_1 - \sigma_2}{\sigma_1 + 2\sigma_2}\cdot\frac{a^3 j_0}{\sigma_2}, \qquad B = -\frac{3j_0}{\sigma_1 + 2\sigma_2}.$$

Thus, we have shown that the field within the sphere is uniform and of intensity $\boldsymbol{E} = 3\boldsymbol{j}_0/(\sigma_1 + 2\sigma_2)$; the associated flow is accordingly also uniform and of density $\boldsymbol{j} = \sigma_1 \boldsymbol{E} = 3\sigma_1 \boldsymbol{j}_0/(\sigma_1 + 2\sigma_2)$.

Calculating the radial component of $\boldsymbol{E}$ just inside and just outside the sphere, we find that

$$E_{1r} = -\partial V_1/\partial r = 3j_0\cos\theta/(\sigma_1 + 2\sigma_2),$$
$$E_{2r} = -\partial V_2/\partial r = 3\sigma_1 j_0\cos\theta/\sigma_2(\sigma_1 + 2\sigma_2).$$

Assuming that the dielectric constants of the conductors are unity, equation (1.20) shows that the surface density of the charge accumulating on the interface is

$$\varepsilon_0(E_{2r} - E_{1r}) = \frac{3(\sigma_1 - \sigma_2)}{\sigma_2(\sigma_1 + 2\sigma_2)}\varepsilon_0 j_0\cos\theta.$$

(We are assuming the equation $\operatorname{div}\boldsymbol{D} = \rho$ remains valid in these circumstances; see § 5.6.) □

All the electrostatic problems solved in Chapters 1 and 2 can be restated as current flow problems and their solutions remain unaltered in form. The only essentially new feature by which current flow problems differ from those relating to electrostatic fields in dielectrics, is the boundary condition $j_n = 0$ over the insulated surface of a conductor. We shall accordingly close this section by giving two examples of problems in which this condition arises and which are solved by introducing image electrodes.

Problem 3.8 A semi-infinite solid of conductivity σ is bounded by a plane face. A pair of small spherical electrodes of radius δ are set into the solid a distance $2a$ apart and at the same distance a from the plane face. Calculate the equivalent resistance between the electrodes.

Solution. Suppose that a current i enters the conductor via electrode A and leaves via electrode B (Fig. 3.4). Imagining the whole of space to be filled with the conducting medium, we introduce image electrodes of strengths $+i$ and $-i$ at A′ and B′ respectively, A′ being the optical image of A in the boundary plane and B′ being the optical image of B. At any point P on the boundary plane it follows from the symmetry of the electrode system with respect to this plane that the resultant current density vector of the flow generated by the four electrodes is parallel to the plane. Hence, if $\boldsymbol{n}$ is the unit normal to the plane, $\boldsymbol{j} \cdot \boldsymbol{n} = 0$ and the boundary condition is satisfied. We conclude that the flow generated by the four electrodes is identical with the actual flow in the semi-infinite region occupied by the conductor.

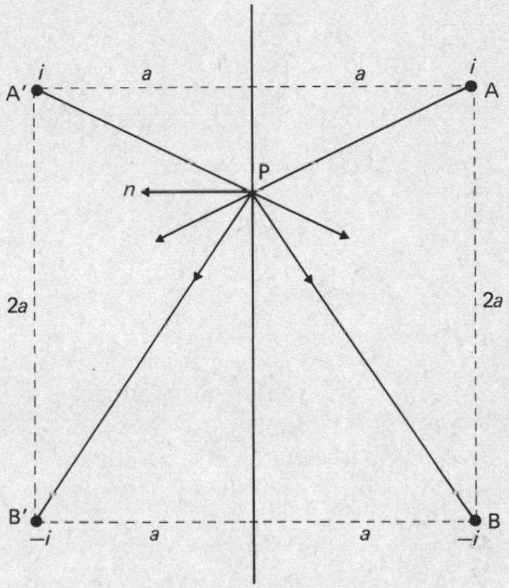

Fig. 3.4

Employing equation (3.10) the potentials at the surfaces of the electrodes A and B, can now be found; they are given by

$$V_{\mathrm{A}} = \frac{i}{4\pi\sigma}\left[\frac{1}{\delta}+\frac{1}{2a}-\frac{1}{2a}-\frac{1}{2\sqrt{2}a}\right] = \frac{i}{4\pi\sigma}\left[\frac{1}{\delta}-\frac{1}{2\sqrt{2}a}\right],$$

$$V_{\mathrm{B}} = \frac{i}{4\pi\sigma}\left[-\frac{1}{\delta}-\frac{1}{2a}+\frac{1}{2a}+\frac{1}{2\sqrt{2}a}\right] = \frac{i}{4\pi\sigma}\left[-\frac{1}{\delta}+\frac{1}{2\sqrt{2}a}\right].$$

The equivalent resistance between the electrodes is therefore

$$\frac{V_{\mathrm{A}}-V_{\mathrm{B}}}{i} = \frac{1}{2\pi\sigma}\left[\frac{1}{\delta}-\frac{1}{2\sqrt{2}a}\right].$$ □

Problem 3.9 Electrodes of small radius δ are set into a conducting circular disc of radius a, thickness t and conductivity σ. The electrodes lie on the same diameter at equal distances f from the centre. Find the equivalent resistance between them.

Solution. Suppose a current i enters the disc at A and leaves at B (Fig. 3.5). Imagine the disc material is extended to infinity and that additional electrodes of strengths i and $-i$ are introduced at the points A′, B′, which

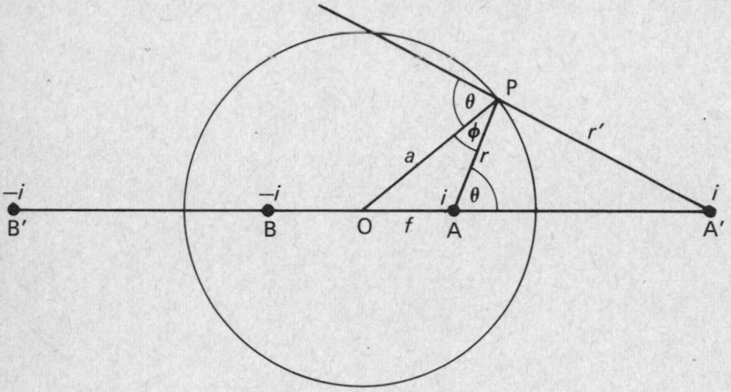

Fig. 3.5

are the inverses of A, B in the circle. The resultant current density at a point P on the edge of the disc due to the two electrodes at A and A′ is found by the use of equation (3.11) to have a component in the direction of the radius OP equal to

$$\frac{i}{2\pi tr}\cos\phi - \frac{i}{2\pi tr'}\cos\theta,$$

where $r = \text{AP}$, $r' = \text{A'P}$. Putting $r' = ar/f$, $\cos\phi = (a^2+r^2-f^2)/2ar$, $\cos\theta = (a^2-r^2-f^2)/2rf$, in this expression, it will be found to reduce to $i/2\pi ta$. An exactly similar argument shows that the same component of the current density generated by the electrodes at B, B′ has value $-i/2ta$. Thus, the flow generated by all four electrodes satisfies the boundary condition $j_n = 0$ at the edge of the disc and accordingly represents the actual flow in the disc.

The potential of the electrode A is now calculated (equation (3.11)) to be

$$\begin{aligned} V_A &= \frac{i}{2\pi\sigma t}\left[-\log\delta + \log 2f - \log\left(\frac{a^2}{f} - f\right) + \log\left(\frac{a^2}{f} + f\right)\right] \\ &= \frac{i}{2\pi\sigma t}\log\left[\frac{2f(a^2+f^2)}{\delta(a^2-f^2)}\right]. \end{aligned}$$

The potential V_B of the electrode B is the negative of this. The equivalent resistance between the electrodes is accordingly

$$\frac{V_A - V_B}{i} = \frac{1}{\pi\sigma t}\log\left[\frac{2f(a^2+f^2)}{\delta(a^2-f^2)}\right]. \qquad \square$$

EXERCISES

1. Conducting wires, all of equal resistance r, form the edges of a cube. Prove that the equivalent resistance between two adjacent corners is $7r/12$.

2. Five small circular electrodes, each of radius δ, are set into an infinite plane sheet of conductivity σ and thickness t, at the vertices and centre of a square of side a. A steady current $\frac{1}{4}i$ enters the sheet at each of the vertices and a current i leaves at the centre. Show that the equivalent resistance between the four vertices and the centre is

$$\frac{5}{8\pi\sigma t}\log\left(\frac{a}{2^{0\cdot 9}\delta}\right).$$

3. A perfectly conducting sphere of radius a is surrounded by a concentric spherical shell of inner radius a and thickness λa, and of conductivity σ; the rest of space is filled by a medium of conductivity σ'. The conducting sphere is maintained at a potential ϕ above that of points at an infinite distance. Show that the current flowing from the sphere is

$$4\pi a\sigma\sigma'(\lambda+1)\phi/(\sigma+\lambda\sigma').$$

4. A uniform conducting medium extends to infinity in all directions and has conductivity σ_0. It supports a uniform flow whose density is $\boldsymbol{j}_0$. A spherical cavity of radius a is cut in the medium and is filled with material whose conductivity at distance r from the centre is $\sigma_0 a/r$. Assuming that the potential within the sphere is of the form $V = \psi(r)\cos\theta$, where r, θ are spherical polar coordinates, show that the ratio of the total current through the sphere to the current which flowed through the same region before the cavity was cut, is $3/(\sqrt{2}+1)$.

5. Two small circular electrodes, each of radius δ, are a distance $2f$ apart in an infinite plane sheet of thickness t and conductivity σ. A circular hole of radius $a(<f)$, whose centre is midway between the electrodes, is punched in the sheet. A current i enters at one electrode A and leaves at the other B. By placing image electrodes of strengths $+i$ and $-i$ at the points inverse to A and B in the circle, show that the effective resistance

between the electrodes is

$$\frac{1}{\pi\sigma t}\log\frac{2f(f^2+a^2)}{\delta(f^2-a^2)}.$$

6. Current flows between two perfectly conducting circular electrodes of radius a in an infinite plane sheet of conductivity σ and thickness t. If their centres are a distance $b(> 2a)$ apart, show that the resistance between them is

$$-\frac{1}{\pi\sigma t}\log\left[\frac{b}{2a}-\left(\frac{b^2}{4a^2}-1\right)^{\frac{1}{2}}\right].$$

(*Hint:* Refer to Problem 1.22.)

Chapter 4

Steady Electromagnetic Fields

4.1 The Biot–Savart Law In this chapter, we shall be concerned with the magnetic field associated with steady currents of the types studied in Chapter 3. The electric field responsible for the flow of charge together with the magnetic field are regarded as two aspects of a single unified field called an *electromagnetic field*. However, in the steady conditions assumed in this chapter, no interaction between the two component fields takes place and they may, accordingly, be studied separately.

The theory will be based upon the *Biot–Savart Law*: Let $\boldsymbol{ds}$ be any vector element of a wire carrying a current i ($\boldsymbol{ds}$ is taken in the same sense as i) and let $\boldsymbol{r}$ be the position vector, relative to the element as origin, of any point P (Fig. 4.1). Then the contribution of the charge flowing in the element to the magnetic intensity at P is given by

$$\boldsymbol{dH} = \frac{i\,\boldsymbol{ds}\times\boldsymbol{r}}{4\pi r^3}. \tag{4.1}$$

This means that $\boldsymbol{dH}$ has magnitude $i\,ds\,\sin\theta/4\pi r^2$, where θ is the angle between $\boldsymbol{ds}$ and $\boldsymbol{r}$, and its direction is that in which P would move if it were rotated in the positive sense (right-hand screw) about an axis along $\boldsymbol{ds}$.

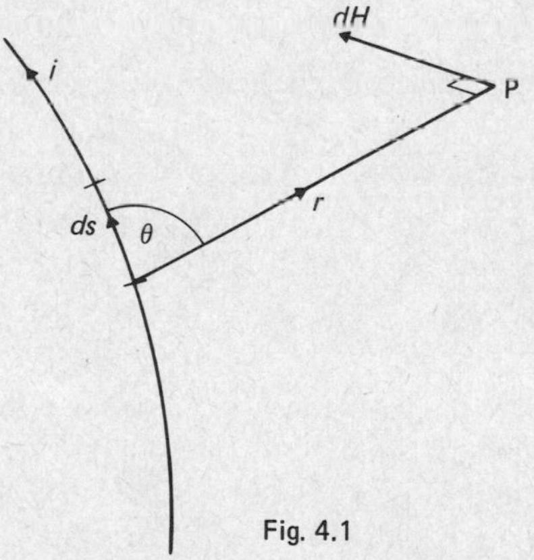

Fig. 4.1

Problem 4.1 Calculate the magnetic intensity at any point on the axis of a circular wire of radius a carrying a current i.

Solution. P is a point on the axis of the wire distant x from its centre O (Fig. 4.2). The contribution to the magnetic intensity at P of the element QQ′ of the wire has magnitude $i\,ds/4\pi r^2$, where ds = QQ′ and QP = r; the direction of this contribution is perpendicular to both QQ′ and QP, i.e. at an angle $\frac{1}{2}\pi - \alpha$ to OP. The axial symmetry indicates that the resultant field at P will be directed along OP and it follows that only the component $i\,ds \sin\alpha/4\pi r^2$ of the contribution of QQ′ need be retained. Integrating

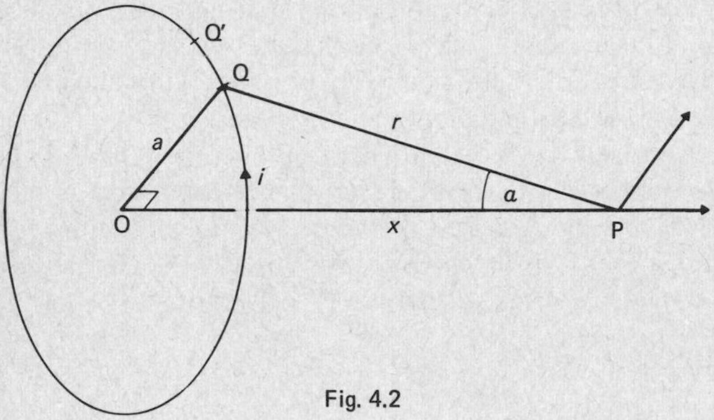

Fig. 4.2

these contributions around the circle, we obtain the net intensity at P as

$$H = \frac{i \sin\alpha}{4\pi r^2}\int ds = \frac{ia\sin\alpha}{2r^2} = \frac{ia^2}{2(a^2+x^2)^{\frac{3}{2}}}. \tag{4.2}$$

In particular, at the centre O the field has intensity $i/2a$. □

Problem 4.2 A steady current i is flowing around a square circuit of side $2a$. Prove that the magnetic intensity at the centre of the square has magnitude $\sqrt{2}i/\pi a$.

Solution. Let QQ′ be an element dx of the side AB of the square distant x from its midpoint (Fig. 4.3). The current i flowing in this element contributes to the field at the centre O an intensity $i\,dx\cos\theta/(4\pi r^2)$ in a direction perpendicular to the plane of the circuit. Putting $x = a\tan\theta$, $dx = a\sec^2\theta d\theta$, $r = a\sec\theta$, this contribution takes the form $i\cos\theta d\theta/(4\pi a)$. By integration over the range $-\frac{1}{4}\pi \leqslant \theta \leqslant \frac{1}{4}\pi$, the net contribution of the side AB is found to be

$$\frac{i}{4\pi a}\int_{-\frac{1}{4}\pi}^{\frac{1}{4}\pi} \cos\theta\, d\theta = \frac{\sqrt{2}i}{(4\pi a)}.$$

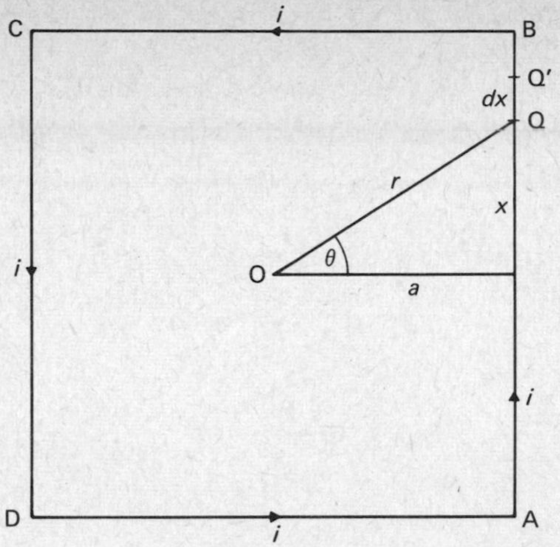

Fig. 4.3

The remaining three sides each make the same contribution, so that the result follows. □

By extending the range of the above integration to $-\frac{1}{2}\pi \leqslant \theta \leqslant \frac{1}{2}\pi$, the intensity at a distance a from a very long straight wire carrying a current i is found to be given by

$$H = \frac{i}{2\pi a}. \tag{4.3}$$

Problem 4.3 Wire is wound closely in a spiral fashion to form a circular disc. Current is sent through the wire so that i is the current across unit length of any radius. Show that the field intensity at any point on the axis of the disc is $\frac{1}{2}i[\cosh^{-1}(\sec\alpha)-\sin\alpha]$, where α is the angle subtended at the point by any radius of the disc.

Solution. The current flowing around a ring-shaped element of the disc (Fig. 4.4) of radius r and width dr is $i\,dr$. It follows from equation (4.2) that the intensity contributed by this element at P on the axis is $\frac{1}{2}ir^2dr(r^2+x^2)^{-\frac{3}{2}}$. Putting $r = x\tan\theta$, $dr = x\sec^2\theta\,d\theta$, this reduces to $\frac{1}{2}i\tan^2\theta\cos\theta\,d\theta$. Hence, the net intensity at P due to the whole disc is

$$\begin{aligned}\tfrac{1}{2}i\int_0^\alpha \tan^2\theta\cos\theta\,d\theta &= \tfrac{1}{2}i\int_0^\alpha (\sec\theta-\cos\theta)\,d\theta \\ &= \tfrac{1}{2}i[\log(\sec\alpha+\tan\alpha)-\sin\alpha].\end{aligned}$$

This is equivalent to the result stated. □

It can be proved that the law (4.1) is equivalent to the statement that the field generated by a circuit C carrying a current i can be derived from a magnetic potential $\Omega = i\omega/4\pi$, where ω is the solid angle subtended at a point of the field by C. Thus

$$\boldsymbol{H} = -\text{grad}\left(\frac{i\omega}{4\pi}\right). \tag{4.4}$$

It should be noted that this potential is not single-valued, since the solid angle ω is arbitrary to the extent of an added multiple of 4π. Thus,

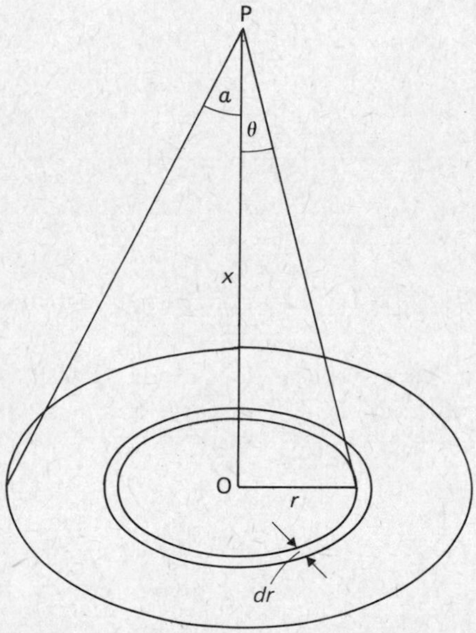

Fig. 4.4

if P is taken around a closed path linking C, the continuous variation of ω will cause its value to change by 4π. The implication is that the field is not conservative; the fact is that such fields can be made to perform work, the energy necessary being supplied by the current source. The operation of the electric motor is based upon this principle.

Problem 4.4 Calculate the magnetic potential due to a long straight wire carrying a current i.

Solution. The potential Ω can be calculated by use of the result (4.4), but it is more easily derived from equation (4.3).

Since the field is two-dimensional, Ω will depend on the cylindrical polar coordinates (ρ, ϕ) alone (the wire is being taken as z-axis). The polar components of the field intensity are given by

$$-\frac{\partial \Omega}{\partial \rho} = 0, \qquad -\frac{1}{\rho}\frac{\partial \Omega}{\partial \phi} = \frac{i}{2\pi\rho}.$$

It now follows that Ω is independent of ρ and that

$$\Omega = -\frac{i\phi}{2\pi}. \tag{4.5}$$

Since ϕ is multi-valued, we again note that Ω is multi-valued also. Also, since the reference line $\phi = 0$ can be chosen arbitrarily, this potential is indeterminate in respect of an added constant. □

4.2 Force on a Current If a wire carrying a current i is placed in a magnetic field of induction $\boldsymbol{B}$, each element $\boldsymbol{ds}$ of the wire experiences a force $\boldsymbol{dF}$ given by

$$\boldsymbol{dF} = i\,\boldsymbol{ds} \times \boldsymbol{B}. \tag{4.6}$$

Note that the force is perpendicular to both the current and the field. This formula remains valid if the current is embedded in magnetic material.

A particular case is the force between a pair of long straight parallel wires carrying currents i and i' and distance a apart. Using equation (4.3), the force is found to be $\mu_0\, ii'/2\pi a$ per unit length and is one of attraction if the currents are in the same sense and of repulsion otherwise.

Problem 4.5 A current i is flowing around a circle of radius a. A dipole of moment $\boldsymbol{m}$ is placed on and along the axis of the circle at a point distant c from the centre. Show that the force acting on the circuit is $\frac{3}{2}\mu_0\, ima^2c(a^2+c^2)^{-\frac{5}{2}}$.

Solution. Let $\boldsymbol{ds}$ be an element of the circuit at a point P (Fig. 4.5) and let $\boldsymbol{r}$ be the position vector of P with respect to the dipole at O. The field at P due to the dipole follows from equation (1.12); substituting in equation (4.6), it is found that

$$\boldsymbol{dF} = \frac{\mu_0 i}{4\pi}\left[\frac{3\boldsymbol{m}\,.\,\boldsymbol{r}}{r^5}\boldsymbol{ds} \times \boldsymbol{r} - \frac{\boldsymbol{ds} \times \boldsymbol{m}}{r^3}\right].$$

$\boldsymbol{ds} \times \boldsymbol{m}$ is directed along the radius PC and the corresponding component of $\boldsymbol{dF}$ is cancelled by a similar component acting on the element of the circuit diametrically opposite to $\boldsymbol{ds}$. $\boldsymbol{ds} \times \boldsymbol{r}$ is directed along the the perpendicular to OP lying in the plane OPC and its contribution to $\boldsymbol{dF}$ can be split into two components (i) $r\,ds \cos\theta$ along PC and (ii) $r\,ds \sin\theta$ parallel to OC. The first of these components is again cancelled by a similar component arising from the element opposite P. The second component can be integrated over all elements of the circuit to give a resultant force

$$F = \int \frac{\mu_0 i}{4\pi} \cdot \frac{3m \cos \theta}{r^3} \, ds \sin \theta$$

along OC. Since r, θ do not vary as P describes the circle, this gives $3\mu_0 \, ima \sin \theta \cos \theta / 2r^3$, which is equivalent to the result stated. □

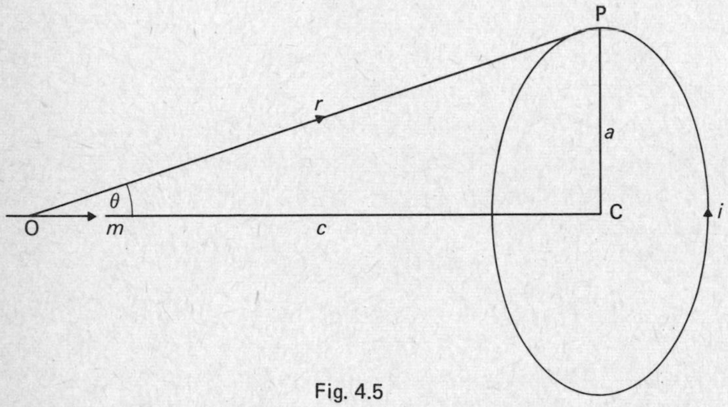

Fig. 4.5

Problem 4.6 Find the force of interaction between a current i flowing in a circular wire of radius a and a current i' flowing in a long straight wire which lies in the plane of the other wire and is distant $c(> a)$ from its centre.

Solution. Consider the element PP′ of the circular wire subtending an angle $d\theta$ at its centre O (Fig. 4.6). The magnetic induction at P due to the current i' has magnitude $\mu_0 i'/2\pi \mathrm{PN} = \mu_0 i'/2\pi(c - a\cos\theta)$ (equation (4.3)) and its direction is perpendicular to the plane of the two wires. The force exerted upon PP′ is therefore $\mu_0 ii'a \, d\theta / 2\pi(c - a\cos\theta)$ in the direction OP. Because of the symmetry about OA, only the component of this force along OA contributes to the resultant force acting upon the circle and this has magnitude $\mu_0 ii'a \cos\theta \, d\theta / 2\pi(c - a\cos\theta)$. Integrating this over the range $0 \leqslant \theta \leqslant 2\pi$, the resultant force of attraction of the circle towards the straight wire is found to be

$$\mu_0 ii' \left[\frac{c}{\sqrt{(c^2 - a^2)}} - 1 \right].$$

(*Note*: The integration can be performed by expressing the integral as a contour integral around the circle $|z| = 1$ in the complex plane, or by changing the variable of integration to $t = \tan\frac{1}{2}\theta$.) □

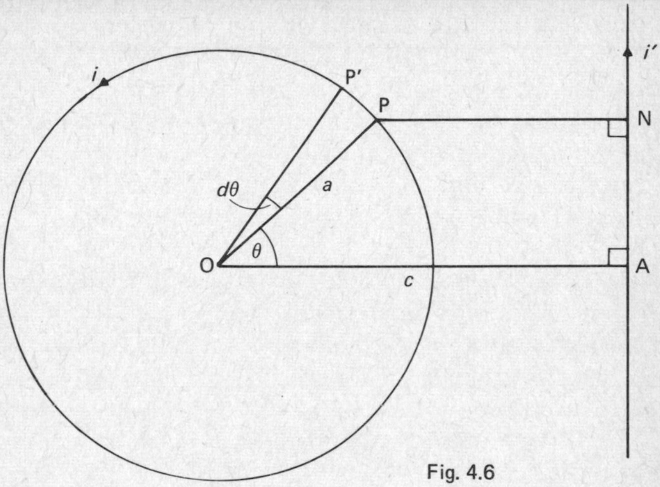

Fig. 4.6

4.3 Potential Energy of a Current Loop If a steady current i is caused to flow in a loop of wire and this is placed in a steady magnetic field whose induction (in the absence of the loop) is $\boldsymbol{B}$, then the potential energy of the system is $-iN$, where N is the flux of $\boldsymbol{B}$ through the loop (more precisely, the flux of $\boldsymbol{B}$ through any surface S bounded by the loop). This statement remains valid if the current is embedded in magnetic material.

If the wire is displaced and the current i is maintained constant, the work done by the forces acting upon the wire will equal the decrease in the potential energy.

***Problem* 4.7** If the current i lies in a plane and encloses an area A and if the field $\boldsymbol{B}$ is uniform, show that the system of forces acting on the loop is equivalent to a couple of moment $iA\boldsymbol{n} \times \boldsymbol{B}$, where $\boldsymbol{n}$ is a unit normal to the plane of the loop.

Solution. Suppose the system of forces acting on the loop is equivalent to a force $\boldsymbol{F}$ acting at some point O in its plane and a couple $\boldsymbol{G}$. If the loop is translated (without rotation) through a small displacement $\boldsymbol{da}$, the work done by the force system is $\boldsymbol{F} . \boldsymbol{da}$. But the flux N of $\boldsymbol{B}$ through the loop remains unchanged and, hence, $\boldsymbol{F} . \boldsymbol{da} = 0$. Since this equation is valid for arbitrary $\boldsymbol{da}$, $\boldsymbol{F}$ vanishes and the force system is equivalent to a couple.

Next suppose the loop is given a small rotation $\boldsymbol{d\theta}$ about O. Then the normal $\boldsymbol{n}$ will undergo this rotation and will change to $\boldsymbol{n}+\boldsymbol{d\theta}\times\boldsymbol{n}$. Thus, the flux of $\boldsymbol{B}$ through the loop will change from $A\boldsymbol{B}.\boldsymbol{n}$ to $A\boldsymbol{B}.(\boldsymbol{n}+\boldsymbol{d\theta}\times\boldsymbol{n})$, i.e. will increase by $A\boldsymbol{B}.\boldsymbol{d\theta}\times\boldsymbol{n}$. This means that the potential energy will decrease by $iA\boldsymbol{B}.\boldsymbol{d\theta}\times\boldsymbol{n} = iA\boldsymbol{n}\times\boldsymbol{B}.\boldsymbol{d\theta}$. But the work done by $\boldsymbol{G}$ is $\boldsymbol{G}.\boldsymbol{d\theta}$ and, hence, $\boldsymbol{G}.\boldsymbol{d\theta} = iA\boldsymbol{n}\times\boldsymbol{B}.\boldsymbol{d\theta}$. This being true for arbitrary $\boldsymbol{d\theta}$, we conclude that $\boldsymbol{G} = iA\boldsymbol{n}\times\boldsymbol{B}$.

It should be noted that this is exactly the couple which the field $\boldsymbol{B}$ would exert upon a magnetic dipole of moment $iA\boldsymbol{n}$. This is an important result; namely, any small current loop i enclosing a plane area dS whose unit normal is $\boldsymbol{n}$ behaves like a magnetic dipole of moment $i\,dS\boldsymbol{n}$. □

4.4 Method of Images If a steady electromagnetic field is generated by electric currents, then over any region devoid of currents, the equations (2.28) or (2.30) which govern a field due to permanent magnets are applicable. Unlike the magnetostatic case, however, the potential Ω will be multi-valued. In particular, within a material of constant permeability, Ω will satisfy Laplace's equation and the method of harmonic functions and images are applicable.

Problem 4.8 Soft iron of permeability μ occupies the whole of space to one side of a plane. A current i flows in a long straight wire which is outside the iron and parallel to its plane face. Show that the wire is attracted towards the iron by a force $\mu_0 i^2(\mu-1)/4\pi a(\mu+1)$ per unit length, a being the distance of the wire from the iron.

Solution. The method is similar to that employed in Problems 2.8 and 2.16. Outside the iron, the field intensity is assumed to be generated by the actual current i through A and a parallel current i' through A′, the optical

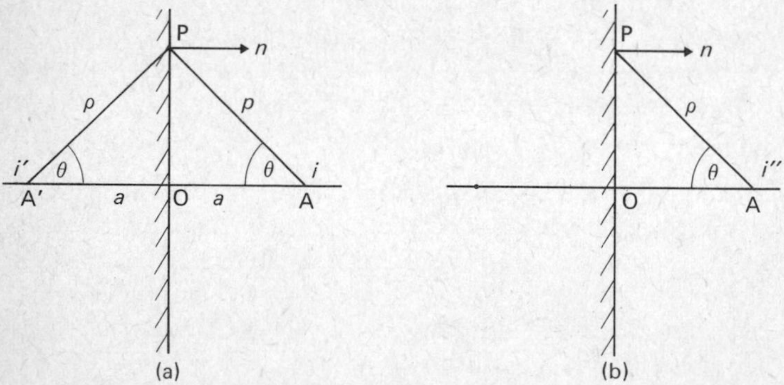

Fig. 4.7

image of A in the plane face (Fig. 4.7a). Inside the iron, the field intensity is supposed generated by a current i'' in the actual wire through A (Fig. 4.7b).

Then, the potential of the field outside the iron at P on the plane face is, by equation (4.5), $[-i(\pi-\theta)-i'\theta]/2\pi$. The potential at P for the field inside the iron is $-i''(\pi-\theta)/2\pi$. Since an arbitrary constant can be added to these functions, the condition that the potential should be continuous across the plane face is

$$i-i' = i''. \tag{4.7}$$

Using equations (4.3) and $\boldsymbol{B} = \mu\mu_0 \boldsymbol{H}$, it is found that the component of $\boldsymbol{B}$ at P normal to the plane face generated by i and i' is given by $B_n = -\mu_0(i+i')\sin\theta/2\pi\rho$. Inside the iron at P, $B_n = -\mu\mu_0 i''\sin\theta/2\pi\rho$. Hence, B_n is continuous across the face provided

$$i+i' = \mu i''. \tag{4.8}$$

Solving equations (4.7) and (4.8) we get $i' = (\mu-1)i/(\mu+1), i'' = 2i/(\mu+1)$. The force acting upon i is that due to the image current i' and, using the result derived on page 67, this is now found to be the result stated. □

Problem 4.9 A long circular cylindrical cavity of radius a is drilled through a large block of iron of permeability μ. A current i flows in a straight wire parallel to the axis of the cavity and returns along a second similar wire. The wires are on opposite sides of, and at the same distance f from, the cavity axis. Calculate the force between them.

Solution. A right-section of the cavity is shown in Fig. 4.8, A and B being the points in which the plane of section is intersected by the two wires. We introduce parallel image currents $\pm i'$ to meet the plane in the points A′, B′ inverse to A, B in the circular boundary of the cavity, and

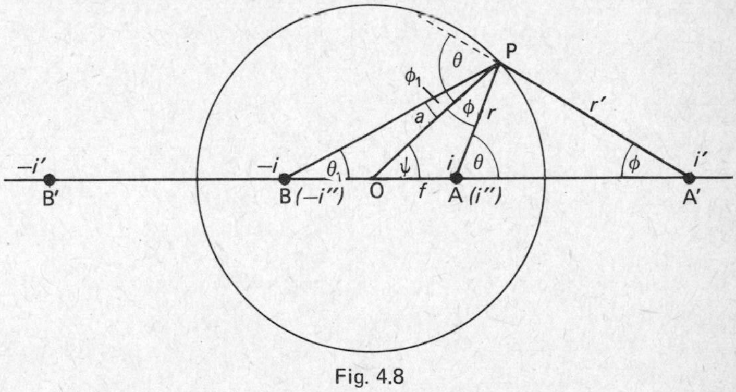

Fig. 4.8

assume that the magnetic intensity in the cavity can be generated by the system of four currents. The intensity in the iron we shall assume can be generated by currents $\pm i''$ at A, B.

Then, the normal component H_n along PO of intensity at P in the cavity generated by the currents at A, A′ is $(i \sin \phi/2\pi r)+(i' \sin \theta/2\pi r')$. Since, by the sine rule, $\sin \phi/r = \sin \theta/r'$, this component reduces to $(i+i')\sin \theta/2\pi r'$. The same component of $\boldsymbol{H}$ in the iron at P generated by i'' at A is $i'' \sin \theta/2\pi r'$. The corresponding contributions of the currents at B, B′ will be $-(i+i')\sin \theta_1/2\pi r_1'$ and $-i''\sin \theta_1/2\pi r_1'$, where θ_1 = angle PBO and $r_1' = \text{B'P}$. It now follows that B_n is continuous at P provided

$$i+i' = \mu i''. \tag{4.9}$$

The magnetic potential at P in the cavity due to i, i' at A, A′ is $-[i\theta+i'(\pi-\phi)]/2\pi = -[i\theta+i'(\pi-\theta+\psi]/2\pi$. Similarly, the potential contributed by $-i$, $-i'$ at B, B′ is $(i\theta_1+i'\phi_1)/2\pi = [i\theta_1+i'(\psi-\theta_1)]/2\pi$. The net potential due to the four currents is accordingly $(i-i')(\theta_1-\theta)/2\pi$ (neglecting an added constant). Similarly, the potential at P in the iron due to the currents $\pm i''$ at A, B is calculated to be $i''(\theta_1-\theta)/2\pi$. The condition for the potential to be continuous at P is therefore

$$i-i' = i''. \tag{4.10}$$

Solving equations (4.9), (4.10), we find that $i' = (\mu-1)i/(\mu+1)$, $i'' = 2i/(\mu+1)$. It now follows that the resultant force acting on unit length of the current i at A due to the currents i' at A′, $-i$ at B and $-i'$ at B′, which is directed along OA, is (see page 67)

$$\frac{\mu_0 i^2}{2\pi}\left[\frac{\mu-1}{\mu+1}\frac{f}{a^2-f^2}+\frac{1}{2f}+\frac{\mu-1}{\mu+1}\frac{f}{a^2+f^2}\right] = \frac{\mu_0 i^2}{2\pi}\left[\frac{1}{2f}+\frac{\mu-1}{\mu+1}\frac{2a^2f}{a^4-f^4}\right]. \quad \square$$

4.5 Ampère's Law If the steady flow of charge is not confined to wires but takes place in a three-dimensional conducting medium and is specified by a current density $\boldsymbol{j}$, the intensity $\boldsymbol{H}$ of the magnetic field generated is given by the equation

$$\operatorname{curl} \boldsymbol{H} = \boldsymbol{j}. \tag{4.11}$$

This equation follows from equation (4.4) thus: if C is a closed path embracing the current i and it is described in a positive sense with respect to i, equation (4.4) implies that

$$\oint_C \boldsymbol{H}.d\boldsymbol{r} = -\frac{i}{4\pi}[\omega],$$

where $[\omega]$ is the increment in ω for one circuit of C. But $[\omega] = -4\pi$ and hence

$$\oint_C \boldsymbol{H} \cdot d\boldsymbol{r} = i.$$

This is *Ampère's Law*. Assuming this law to be valid for the steady flow $\boldsymbol{j}$, since the current through C will be given by the flux of $\boldsymbol{j}$ across a surface S bounded by C, we have

$$\oint_C \boldsymbol{H} \cdot d\boldsymbol{r} = \int_S \boldsymbol{j} \cdot \boldsymbol{n}\, dS. \tag{4.13}$$

Application of Stokes' Theorem (L. Marder, *Vector Fields*, p. 50) now leads to equation (4.11).

To summarize, the equations of the magnetic field in a medium of permeability μ due to a given current distribution $\boldsymbol{j}$ are

$$\operatorname{curl} \boldsymbol{H} = \boldsymbol{j}, \qquad \operatorname{div} \boldsymbol{B} = 0, \qquad \boldsymbol{B} = \mu\mu_0 \boldsymbol{H}. \tag{4.14}$$

In regions where $\boldsymbol{j}$ is zero, a potential Ω exists and

$$\boldsymbol{H} = -\operatorname{grad} \Omega. \tag{4.15}$$

Problem 4.10 An infinitely long straight rod has a circular cross-section of radius a. A current i flows in the rod, its distribution over any cross-section being uniform. Calculate the magnetic field generated.

Solution. Taking axes $Oxyz$, with Oz along the axis of the rod, and introducing cylindrical polar coordinates (ρ, ϕ, z) in the usual way, let (B_ρ, B_ϕ, B_z) be the corresponding radial, transverse and z-components of $\boldsymbol{B}$. The axial symmetry requires that all these components shall be functions of ρ alone. Hence, constructing a right circular cylinder, coaxial with the rod, of radius ρ and unit length, the fluxes of $\boldsymbol{B}$ out of the two circular ends will cancel; also, the flux of $\boldsymbol{B}$ out of the curved surface will be $2\pi\rho B_\rho$. But the flux of $\boldsymbol{B}$ out of any closed surface always vanishes and, hence, $B_\rho = 0$. Thus $H_\rho = 0$.

The current through one of the circular ends of the cylinder is i if $\rho > a$ and is $i\rho^2/a^2$ if $\rho < a$. Thus, if H_ϕ is the component of $\boldsymbol{H}$ tangential to the circumference of this end, then H_ϕ is constant and Ampère's law applied to this circle gives

$$\begin{aligned} 2\pi\rho H_\phi &= i\rho^2/a^2, \quad \rho < a, \\ &= i, \qquad\quad \rho > a. \end{aligned}$$

Hence, inside the rod $H_\phi = i\rho/2\pi a^2$, and outside, $H_\phi = i/2\pi\rho$.

It follows from the previous argument that H_x, H_y, H_z are functions of x, y alone. Hence, the x- and y-components of $\operatorname{curl} \boldsymbol{H}$ are $\partial H_z/\partial y$ and $-\partial H_z/\partial x$ respectively. But, equation (4.11) requires that these components should vanish and, therefore, H_z is constant. Since $\boldsymbol{H}$ must be zero at an infinite distance, this means that $H_z = 0$.

We have proved, therefore, that the field outside the rod is identical with that due to a current i flowing along its axis. □

Problem 4.11 A current of constant density flows in an infinite, plane, thin conducting sheet, the lines of flow being parallel. Calculate the field outside the sheet.

Solution. Take axes $Oxyz$ so that the x-axis lies along a line of flow and the z-axis is normal to the sheet. Then $\boldsymbol{H}$ can only depend upon z and the components of $\operatorname{curl}\boldsymbol{H}$ are therefore $(-\partial H_y/\partial z, \partial H_x/\partial z, 0)$. Since $\operatorname{curl}\boldsymbol{H}$ vanishes outside the sheet, H_x and H_y must both be constant. Hence, $\operatorname{div}\boldsymbol{B} = \partial B_z/\partial z = 0$ and, thus, B_z is also constant. In the region $z < 0$, suppose $B_z = B_0$; then, in the region $z > 0$, $B_z = -B_0$. But the normal component B_z is continuous across the sheet and it follows that $B_0 = 0$. Thus $H_z = 0$.

Now construct a rectangular circuit in the yz-plane whose sides are the lines $y = \pm y_0$, $z = \pm z_0$ and apply Ampère's law. If j is the current density in the sheet and t is its thickness, the current flowing through the circuit is $2jty_0$. The tangential line integral of $\boldsymbol{H}$ around the circuit is $2y_0(H_{1y} - H_{2y})$, where $H_y = H_{1y}$ for $z < 0$ and $H_y = H_{2y}$ for $z > 0$. Ampère's law shows that $H_{1y} - H_{2y} = jt$. But H_y must have the same magnitude in the two regions and, hence, $H_{1y} = \frac{1}{2}jt$ and $H_{2y} = -\frac{1}{2}jt$.

By a similar application of the law to a circuit in the xz-plane whose sides are the lines $x = \pm x_0$, $z = \pm z_0$, we prove that $H_{1x} = H_{2x} = 0$ (there is no current flow through this rectangle).

We have shown, therefore, that the magnetic field in the region $z < 0$ is uniform, of magnitude $\frac{1}{2}jt$ and parallel to Oy. The field in the region $z > 0$ is the same, except that it is in the reverse sense. □

EXERCISES

1. A current circulates around a wire bent into the shape of a semi-circle with its diameter. Show that at a point on a line through the centre perpendicular to the plane of the semi-circle, the intensity makes an angle ϕ with the line given by $\tan\phi = (2\tan\alpha)/\pi$, where 2α is the angle subtended by the diameter at the point.

2. A wire carrying a current i is in the shape of the plane curve having polar equation $r = f(\theta)$. Show that the magnetic intensity at the pole has magnitude

$$\frac{i}{4\pi}\int\frac{d\theta}{r}.$$

Deduce that the field at the focus of an elliptical current has intensity i/l, where l is the latus rectum.

3. A current i is flowing in a square circuit having side of length a. A side of the square subtends an angle 2α at a point P on the perpendicular to the plane of the square through its centre. Show that the magnetic intensity at P has magnitude $(4i \sin \alpha \tan^2\alpha)/\pi a$.

4. A current i flows around a circle of radius a and another current i' flows in a long straight wire which is perpendicular to the plane of the circle and a perpendicular distance c from its centre. Show that there is a couple tending to bring the two wires into the same plane, of moment $\mu_0 ii'a^2/2c$ or $\mu_0 ii'c/2$, according as $c > a$ or $c < a$.

5. A current i flowing in an infinite, straight wire is at a distance f from and parallel to the axis of an infinite circular cylinder of iron of permeability μ. The radius of the cylinder is $a(<f)$. Show that the presence of the iron reduces the magnetic intensity over the region it occupies by a factor $2/(\mu+1)$ and that the wire is attracted to the iron with a force

$$\frac{\mu_0 a^2 i^2(\mu-1)}{2\pi f(f^2-a^2)(\mu+1)}$$

per unit length.

Chapter 5

General Electromagnetic Field

5.1 Electromagnetic Induction If a loop of wire C is placed in a varying magnetic field, an e.m.f. is induced in the wire and causes a current to flow around it. The e.m.f. V is proportional to the rate of change of the flux of magnetic induction N through the loop; if SI units are employed, we can write

$$V = -\frac{dN}{dt}, \tag{5.1}$$

a result which is called *Faraday's Law*. The negative sign is correct provided the positive senses of the e.m.f. around the circuit and of the flux through the circuit are related by a right-hand screw rule; the existence of this sign is frequently called to attention by reference to a *back* e.m.f.

If S is any surface bounded by C, and $\boldsymbol{n}$ is its unit normal, then

$$N = \int_S \boldsymbol{B}.\boldsymbol{n}\, dS. \tag{5.2}$$

Since $\operatorname{div} \boldsymbol{B} = 0$, N is independent of the form of S.

Thus, if a plane loop of wire enclosing an area A is rotated with angular velocity ω in a uniform magnetic field $\boldsymbol{B}$ perpendicular to the axis of rotation, the flux of induction through the loop when the normal to the plane of the loop makes an angle ωt with $\boldsymbol{B}$ is $N = AB \cos \omega t$. Substituting in equation (5.1), the e.m.f. induced is calculated to be $V = AB\omega \sin \omega t$. An e.m.f. varying with t in this manner is called an *alternating e.m.f.*

***Problem* 5.1** A magnetic dipole of moment m is placed with its axis along the axis of a circular wire of radius a at a distance c from its centre. It is then moved along the axis to this centre. If R is the resistance of the wire and its inductance is negligible, show that a charge

$$\frac{\mu_0 m}{2Ra}\left[\left(\frac{a^2}{a^2+c^2}\right)^{\frac{3}{2}} - 1\right]$$

is caused to circulate.

Solution. At time t during the motion of the dipole, let N be the flux of $\boldsymbol{B}$ due to the magnet through the wire. Then dN/dt is the e.m.f. induced in the wire and, by Ohm's law, the current which flows is $i = dN/R\,dt$.

Thus, the charge which circulates during a time interval (t_1, t_2) is equal to

$$q = \int_{t_1}^{t_2} i\,dt = \frac{1}{R}\int_{t_1}^{t_2} \frac{dN}{dt}\,dt = \frac{1}{R}(N_2 - N_1).$$

Let P be a point in the plane of the wire having spherical polar coordinates (r, θ) relative to the initial position of the magnet as pole (Fig. 5.1).

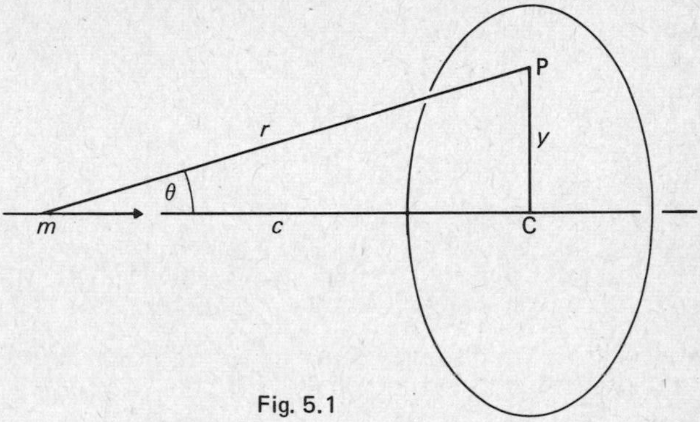

Fig. 5.1

It follows from equations (1.11) (replacing $1/\varepsilon_0$ by μ_0) that the component of $\boldsymbol{B}$ at P normal to the plane of the wire is $\mu_0\, m(3\cos^2\theta - 1)/(4\pi r^3)$. Thus, if CP $= y$, the initial flux of $\boldsymbol{B}$ through the wire is

$$\frac{\mu_0\, m}{4\pi}\int_0^a \frac{1}{r^3}(3\cos^2\theta - 1)\,.\,2\pi y\,dy.$$

Putting $y = c\tan\theta$, $dy = c\sec^2\theta\,d\theta$, $r = c\sec\theta$, this flux is found to be $N_1 = \tfrac{1}{2}\mu_0\, a^2 m(a^2 + c^2)^{-\frac{3}{2}}$. The final flux is found by putting $c = 0$ to be $N_2 = \mu_0\, m/2a$. The value of q now follows. □

5.2 Inductance Between Circuits If the loop C of the last section is situated in a magnetic field generated by a current i' in another loop C', and we assume that the magnetization induced in any materials present is everywhere proportional to $\boldsymbol{H}$ (equation (2.26)), then the induction $\boldsymbol{B}$ at a given point due to the current i' will be proportional to i'. Thus we can write $N = Mi'$. M is called the *mutual inductance* of C and C'. It may be proved that this relationship between C and C' is a reciprocal one and that, if both circuits are immersed in an infinite medium of permeability μ,

$$M = \frac{\mu\mu_0}{4\mu}\int_{C_1}\int_{C_2} \frac{\boldsymbol{ds}_1\,.\,\boldsymbol{ds}_2}{r}, \tag{5.3}$$

where $\boldsymbol{ds}_1$, $\boldsymbol{ds}_2$ are vector elements of the circuits and r is the distance between them.

***Problem* 5.2** Calculate the mutual inductance of the two circuits shown in Fig. 4.6.

Solution. Taking OA as x-axis and the perpendicular to this line through O in the direction of the current i' as y-axis, the induction due to this current at a point (x, y) is of magnitude $B = \mu_0 i'/2\pi(c-x)$. $\boldsymbol{B}$ is everywhere perpendicular to the plane of the circular circuit. Integrating $\boldsymbol{B}$ over the area enclosed by the circle, we obtain for the flux of induction through this circuit,

$$N = \frac{\mu_0 i'}{\pi}\int_{-a}^{a} dx \int_{0}^{\sqrt{(a^2-x^2)}} \frac{dy}{c-x} = \frac{\mu_0 i'}{\pi}\int_{-a}^{a} \frac{\sqrt{(a^2-x^2)}}{c-x}\,dx.$$

It can be verified by differentiation that

$$\int \frac{\sqrt{(a^2-x^2)}}{c-x}\,dx = c\sin^{-1}\frac{x}{a} - \sqrt{(c^2-a^2)}\sin^{-1}\frac{cx-a^2}{a(c-x)} - \sqrt{(a^2-x^2)}.$$

Thus, $N = \mu_0 i'[c-\sqrt{(c^2-a^2)}]$ and the mutual inductance is $\mu_0[c-\sqrt{(c^2-a^2)}]$. □

***Problem* 5.3** Two equal, circular loops of radius a, lie opposite each other, a distance c apart. Show that their mutual inductance is

$$M = \tfrac{1}{2}a^2\mu_0 \int_0^{2\pi} \frac{\cos\theta\, d\theta}{(c^2+2a^2-2a^2\cos\theta)^{\frac{1}{2}}}.$$

If c is large, deduce that $M = \mu_0 \pi a^4/2c^3$.

Solution. Elements $\boldsymbol{ds}_1$, $\boldsymbol{ds}_2$ of the two circuits are shown in Fig. 5.2 as PP′, QQ′ respectively. Taking their scalar product, we find

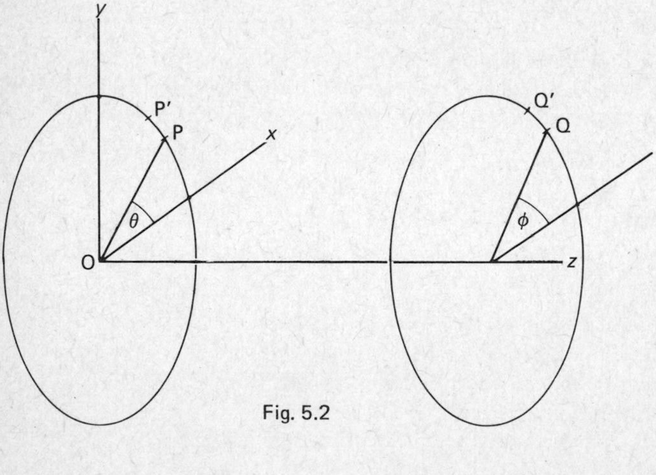

Fig. 5.2

$ds_1 . ds_2 = a^2\cos(\theta-\phi)\, d\theta d\phi$. Relative to the axes $Oxyz$, the coordinates of P are $(a\cos\theta, a\sin\theta, 0)$ and of Q are $(a\cos\phi, a\sin\phi, c)$. Hence

$$PQ = r = [c^2+2a^2-2a^2\cos(\theta-\phi)]^{\frac{1}{2}}.$$

Equation (5.3) now leads to the result

$$M = \frac{\mu_0}{4\pi}\int_0^{2\pi}\int_0^{2\pi}\frac{a^2\cos(\theta-\phi)}{[c^2+2a^2-2a^2\cos(\theta-\phi)]^{\frac{1}{2}}}\, d\theta\, d\phi.$$

But

$$\int_0^{2\pi}\frac{\cos(\theta-\phi)\, d\theta}{[c^2+2a^2-2a^2\cos(\theta-\phi)]^{\frac{1}{2}}} = \int_0^{2\pi}\frac{\cos\theta\, d\theta}{(c^2+2a^2-2a^2\cos\theta)^{\frac{1}{2}}}.$$

This integral is independent of ϕ; the integration with respect to ϕ accordingly simply introduces a factor 2π and the result stated then follows.

If a/c is small, $(c^2+2a^2-2a^2\cos\theta)^{-\frac{1}{2}} = c^{-1}[1-a^2(1-\cos\theta)/c^2]$ to order $(a/c)^2$. The integral upon which M depends has therefore the approximate value $\pi a^2/c^3$. Hence, $M = \mu_0\, \pi a^4/2c^3$. □

5.3 Circuits with Discrete Elements In this section, we shall study electric circuits in which elements called resistors, capacitors and inductors are joined together by wires having negligible resistance.

Resistors and capacitors have been sufficiently described in earlier chapters. An *inductor* is usually a coil of wire wound on a cylindrical former of magnetic material. Suppose a current i is flowing through the coil; assuming that the magnetic induction generated is everywhere proportional to i, the flux of $\boldsymbol{B}$ through each turn of the coil will also be proportional to i. Thus, the total flux of induction through the coil can be taken to be Li, where L is a constant for the element called its *self-inductance*. By Faraday's law, if i varies an e.m.f. V will be induced in the coil given by the equation.

$$V = -L\frac{di}{dt}. \tag{5.4}$$

The negative sign indicates that this e.m.f. acts so as to oppose the change in the current.

If the resistance of an inductor cannot be neglected, this will be represented in a circuit diagram by a separate resistor in series with the coil.

Problem 5.4 An inductor having self-inductance L and resistance R is connected in parallel with a capacitor C in series with a resistor R'. If the combination is connected to a source of constant e.m.f. E, find the conditions satisfied by L, C, R, R' if the current taken is constant.

Solution. The circuit is shown in Fig. 5.3. Let i_1, i_2 be the currents flowing in the two branches at time t after connecting the supply and let q, $-q$ be the charges on the plates of the capacitor at this instant. Then, the drops

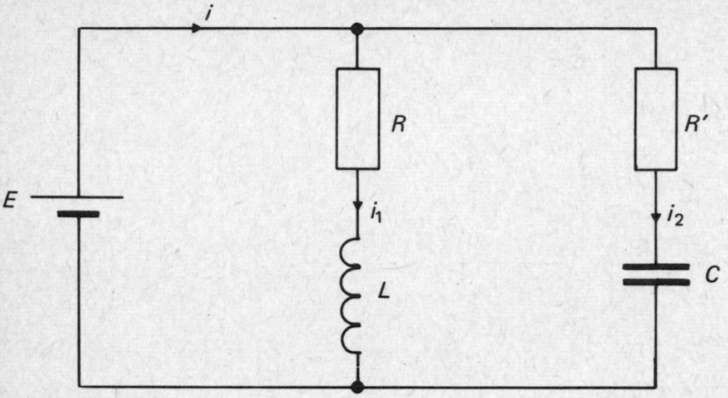

Fig. 5.3

in potential between the ends of the branches are $Ri_1 + L\,di_1/dt$ and $R'i_2 + q/C$. These must both equal the e.m.f. applied by the supply; hence, since $dq/dt = i_2$, the following equations are valid:

$$E = Ri_1 + L\frac{di_1}{dt} = R'\frac{dq}{dt} + \frac{q}{C}. \tag{5.5}$$

Solving for i_1 and q under the initial conditions $i_1 = 0$, $q = 0$ at $t = 0$, we find that

$$i_1 = \frac{E}{R}\left[1 - \exp\left(-\frac{Rt}{L}\right)\right], \quad q = EC\left[1 - \exp\left(-\frac{t}{CR'}\right)\right]. \tag{5.6}$$

Hence $i_2 = dq/dt = (E/R')\exp(-t/CR')$.

If $i = i_1 + i_2$ is to be constant, the two exponential terms must cancel. The conditions for this are $R = R'$ and $R/L = 1/CR'$, i.e. $L = CR^2$. The constant current flowing is then E/R. □

Problem 5.5 A pair of identical circuits, each containing a capacitor C and a coil of inductance L (negligible resistance) in series, are so arranged that the coils are magnetically coupled and have mutual inductance M. At $t = 0$, the currents in both circuits are zero and the capacitors have charges Q and 0. Determine the charges at time $t > 0$. (Assume $M < L$.)

Solution. Employing the notation of Fig. 5.4, and equating the potential drops around the circuits in the senses of the currents to the e.m.f.s in the circuits, we get the equations

$$\frac{q_1}{C} = -LDi_1 - MDi_2, \qquad \frac{q_2}{C} = -LDi_2 - MDi_1.$$

Since $i_1 = Dq_1$, $i_2 = Dq_2$, these equations are equivalent to

$$\begin{aligned}(LD^2+1/C)q_1 + MD^2q_2 &= 0,\\ MD^2q_1 + (LD^2+1/C)q_2 &= 0.\end{aligned} \tag{5.7}$$

Eliminating q_2, we arrive at the following equation for q_1: $[(L-M)D^2+1/C][(L+M)D^2+1/C]q_1 = 0$. This equation has characteristic roots $\pm j\lambda$, $\pm j\mu$, where $\lambda^2 = 1/C(L-M)$, $\mu^2 = 1/C(L+M)$, and the general solution for q_1 can accordingly be written,

$$q_1 = P\cos\lambda t + P'\sin\lambda t + R\cos\mu t + R'\sin\mu t.$$

Substituting this in the first of equations (5.7) and integrating twice, we find that $q_2 = -P\cos\lambda t - P'\sin\lambda t + R\cos\mu t + R'\sin\mu t$.

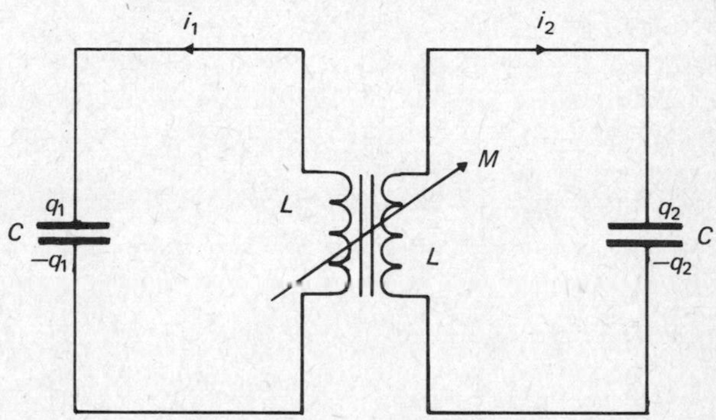

Fig. 5.4

The initial conditions are $t = 0$, $q_1 = Q$, $q_2 = 0$, $Dq_1 = Dq_2 = 0$. To satisfy these, it is necessary that $P = R = \frac{1}{2}Q$, $P' = R' = 0$; thus, $q_1 = \frac{1}{2}Q(\cos\mu t + \cos\lambda t)$, $q_2 = \frac{1}{2}Q(\cos\mu t - \cos\lambda t)$. This is the solution required. It can also be written in the form

$$q_1 = Q\cos\tfrac{1}{2}(\lambda-\mu)t\cos\tfrac{1}{2}(\lambda+\mu)t, \quad q_2 = Q\sin\tfrac{1}{2}(\lambda-\mu)t\sin\tfrac{1}{2}(\lambda+\mu)t.$$

Then, if M is small by comparison with L, λ and μ differ by a small quantity only and both q_1, q_2 execute oscillations of period $4\pi/(\lambda+\mu) \doteqdot 2\pi/\lambda$ with slowly varying amplitudes $Q\cos\frac{1}{2}(\lambda-\mu)t$, $Q\sin\frac{1}{2}(\lambda-\mu)t$. The oscillations of the amplitudes are 90° out of phase with one another, so that when one is at a maximum, the other is zero. This means that energy will be transferred rhythmically between the circuits. This phenomenon is a characteristic feature of coupled oscillatory systems and is also exhibited by two simple pendulums whose points of suspension are mechanically connected. Of course, in any actual system of this type, the circuits will not be entirely devoid of resistance and the oscillations will be damped and will ultimately die away. □

5.4 Alternating Currents In this section, circuits in which the applied e.m.f. varies sinusoidally with the time will be studied. An e.m.f. of this type is supplied by an alternator, in which a coil is rotated in a magnetic field and its magnitude at time t is given by an expression of the form $E_0 \cos(\omega t + \alpha)$. E_0 is the *amplitude*, ω is the *angular frequency* ($\omega/2\pi$ is the frequency in cycles per second or Hertz) and α is the *phase*. It will be assumed that the circuit has reached a steady state in which the currents flowing are also oscillating sinusoidally with the same frequency as the applied e.m.f. (i.e. transients are neglected).

It is convenient to regard the applied e.m.f. as the real part of the complex expression $E_0 \exp[j(\omega t + \alpha)] = E \exp(j\omega t)$, where $E = E_0 \exp(j\alpha)$. E is termed the *complex e.m.f.* Note that $|E| = E_0$ and $\arg E = \alpha$. The currents are treated in the same way, a complex current I being associated with an actual current $\mathcal{R} I \exp(j\omega t)$.

If I is the complex current through an inductance L, the actual potential drop across the element is

$$L\frac{d}{dt}(\mathcal{R} I e^{j\omega t}) = \mathcal{R}\, j\omega L I e^{j\omega t}.$$

Thus, the complex potential drop is $j\omega LI$. $j\omega L$ is called the *complex impedance* of the element at the angular frequency ω. Hence, if V is the complex potential drop, Z is the complex impedance and I is the complex current, then

$$V = ZI. \tag{5.8}$$

(c.f. Ohm's Law, § 3.1). A similar argument proves that equation (5.8) is valid for a resistance R and a capacitance C; the respective complex impedances are R and $1/j\omega C$.

A.C. circuit problems can now be solved by application of Kirchhoff's two laws in the manner explained in § 3.1. The currents flowing are represented by complex quantities and the potential drops are calculated from the complex impedances of the elements by use of equation (5.8). In the solutions of the problems which follow, the complex currents, complex impedances, etc., will be referred to simply as currents, impedances, etc.

Problem 5.6 A coil having resistance R and inductance L, and a condenser having capacitance C, are connected in parallel across an alternator supplying an e.m.f. $E \cos \omega t$. Determine the frequency at which the alternator current is a minimum.

Solution. Let I_1, I_2 be the currents supplied to the coil and condenser respectively (Fig. 5.5). Since the impedances of these elements are $R+j\omega L$,

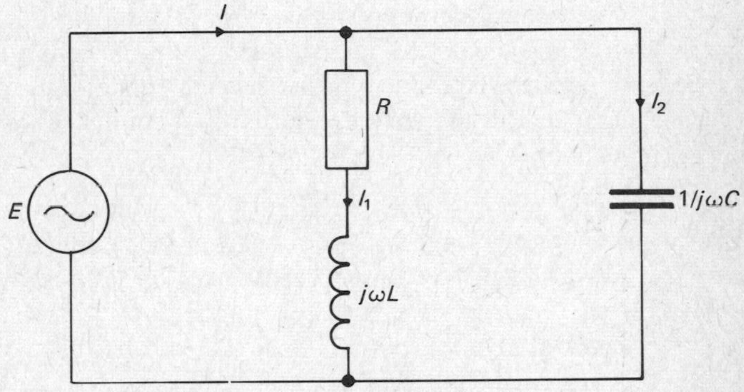

Fig. 5.5

$1/j\omega C$ respectively, it follows that $E = (R+j\omega L)I_1 = I_2/j\omega C$. Hence, if I is the current supplied by the alternator,

$$I = I_1 + I_2 = E\frac{1-\omega^2 LC + j\omega CR}{R+j\omega L}.$$

The amplitude of the actual current is accordingly

$$|I| = E\left[\frac{(1-\omega^2 LC)^2+\omega^2 C^2 R^2}{R^2+\omega^2 L^2}\right]^{\frac{1}{2}}$$

and its phase is

$$\arg I = \arg(1-\omega^2 LC + j\omega CR) - \arg(R+j\omega L)$$

$$= \tan^{-1}\left(\frac{\omega CR}{1-\omega^2 LC}\right) - \tan^{-1}\left(\frac{\omega L}{R}\right).$$

Differentiating $|I|^2$ with respect to ω^2, it will be found that a minimum exists for

$$\omega^2 = \frac{1}{LC}[(1+2\alpha)^{\frac{1}{2}} - \alpha],$$

where $\alpha = CR^2/I = 1/Q^2$ (Q is the *quality factor*). At this frequency, the system is said to be in a state of *amplitude resonance*.

The frequency at which the current I is in phase with the applied e.m.f. E is determined by the equation

$$\frac{\omega CR}{1-\omega^2 LC} = \frac{\omega L}{R}:$$

i.e. $\omega^2 = (1-\alpha)/LC$. This is the state of *phase resonance*. If α is small, the two states of resonance are indistinguishable. □

Problem 5.7 Three-phase a.c. supply of voltage amplitude E is fed to the vertices of a triangle. Calculate the current taken from the supply (i) when these vertices are connected to one another by elements of impedance Z, and (ii) when these vertices are each connected to a common point O by elements of impedance Z.

Solution. Three-phase a.c. supply is carried by three conductors and is such that the alternating voltage across any pair of the conductors differs in phase from the voltage across any other pair by $2\pi/3$. Thus, if A, B, C are the vertices of the triangle connected to the supply, the complex electromotive forces applied across BC, CA, AB can be taken to be E, $E\exp(-2j\pi/3)$, $E\exp(2j\pi/3)$, respectively.

In case (i), the circuit is as shown in Fig. 5.6(i); this is called a *delta connection* of the elements Z to the supply. It follows from the symmetry of the system that the currents in the arms BC, CA, AB can be assumed to

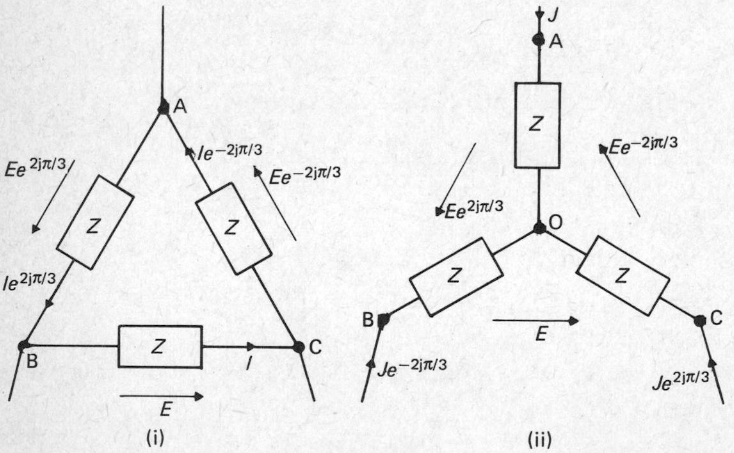

Fig. 5.6

be I, $I\exp(-2j\pi/3)$, $I\exp(2j\pi/3)$ respectively. Since the potential drop across BC is E, it follows that $E = ZI$. The current taken from the supply at A is, therefore,

$$I\exp\frac{2j\pi}{3} - I\exp\frac{-2j\pi}{3} = 2jI\sin\frac{2\pi}{3} = \frac{\sqrt{3}jE}{Z}.$$

This current has amplitude $\sqrt{3}E/|Z|$; the currents taken at B, C have the same amplitude, but are out of phase by an angle $2\pi/3$.

In case (ii), the circuit is as shown in Fig. 5.6(ii); this is called a *star connection* of the elements Z. The currents in AO, BO, CO are taken to be J, $J\exp(-2j\pi/3)$, $J\exp(2j\pi/3)$ respectively. By consideration of the potential drops along BOC, we arrive at the equation

$$E = ZJ[\exp(-2j\pi/3) - \exp(2j\pi/3)] = -\sqrt{3}jZJ.$$

Thus, $J = jE/\sqrt{3}Z$ is the current taken from the supply at A. The amplitudes of the currents taken at A, B, C are all $E/\sqrt{3}|Z|$ and are mutually out of phase by an angle $2\pi/3$. □

***Problem* 5.8** In the telephone circuit shown in Fig. 5.7, the inductance of the telephone windings is included in L' and all resistances are negligible. The e.m.f. supplied is $E\cos\omega t$. Calculate the current through the telephone and the frequency at which there is no signal.

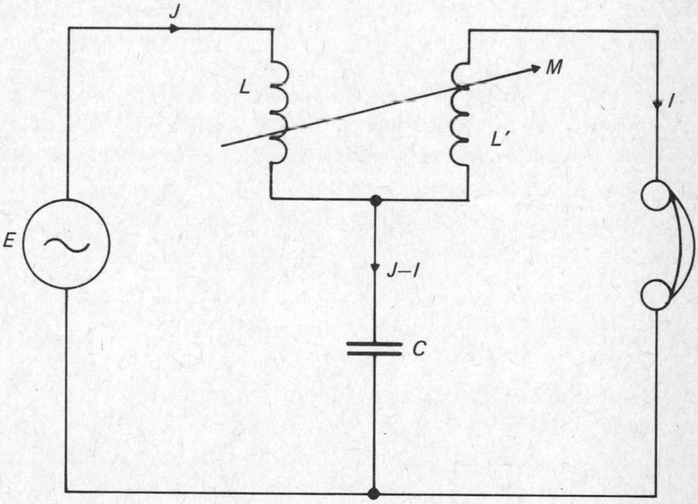

Fig. 5.7

Solution. Let I be the current through the telephone and J the current taken from the supply. Applying Kirchhoff's second law to the left-hand and right-hand circuits, we derive the equations

$$E = j\omega LJ + j\omega MI + (J-I)/j\omega C,$$
$$0 = j\omega L'I + j\omega MJ - (J-I)/j\omega C.$$

Solving for I, we obtain

$$I = \frac{(1+\omega^2 MC)E}{j\omega[L+L'+2M-\omega^2 C(LL'-M^2)]}.$$

Clearly, I vanishes when $\omega^2 = -1/MC$; ω^2 must, of course, be positive and the senses of the windings of the coils L, L' must therefore be so arranged that M is negative. □

5.5 The Lorentz Force The force $\boldsymbol{F}$ which acts upon a point charge q moving with velocity $\boldsymbol{v}$ in an electromagnetic field is given by

$$\boldsymbol{F} = q(\boldsymbol{E} + \boldsymbol{v} \times \boldsymbol{B}). \tag{5.9}$$

The definition of the electric intensity $\boldsymbol{E}$ accounts for the component $q\boldsymbol{E}$. The motion of the charge creates an equivalent current element $q\boldsymbol{v}$; substituting this for $i\,d\boldsymbol{s}$ in equation (4.6), we obtain the second component of force $q\boldsymbol{v} \times \boldsymbol{B}$.

$\boldsymbol{F}$ is called the *Lorentz force.* Note that its magnetic component is perpendicular to the direction of motion of q and, hence, does no work.

Problem 5.9 A point charge moves in a uniform magnetic field. Discuss the motion.

Solution. Let Oxyz be stationary rectangular axes and let (x, y, z) be the coordinates of the charge q at time t. Take the magnetic field to be directed along Oz, so that $\boldsymbol{B} = (0, 0, B)$. The velocity of the charge is given by $\boldsymbol{v} = (\dot{x}, \dot{y}, \dot{z})$. Hence, the force acting is given by $\boldsymbol{F} = q(B\dot{y}, -B\dot{x}, 0)$, and, if m is the mass of the charge, its equations of motion are accordingly

$$m\ddot{x} = qB\dot{y}, \qquad m\ddot{y} = -qB\dot{x}, \qquad m\ddot{z} = 0. \tag{5.10}$$

The last of these equations shows that the motion parallel to Oz is uniform. The other equations can be integrated to give $\dot{x} = \omega y$, $\dot{y} = -\omega x$, where $\omega = qB/m$ and the origin O is assumed to be translated appropriately so that the constants of integration vanish. Elimination of y now leads to the equation $\ddot{x} + \omega^2 x = 0$, and hence $x = a\cos(\omega t + \alpha)$. By proper choice of the instant $t = 0$, α can be made zero and then $\dot{x} = \omega y$ shows that $y = -a\sin\omega t$. Thus the equations of the particle's trajectory are

$$x = a\cos\omega t, \qquad y = -a\sin\omega t, \qquad z = bt + c;$$

these are the equations of a circular helix. If the charge is constrained to move in a plane parallel to Oxy, $b = 0$ and its trajectory is a circle; if ω is positive, the charge moves around the circle in the negative (clockwise) sense with angular velocity ω. □

Problem 5.10 The point charge of the last problem is constrained to move in the plane Oxy and, in addition to the magnetic field, is acted upon by an electric field of intensity $E \cos \omega t$ parallel to the x-axis. If the charge is initially at rest at O, show that its trajectory is a spiral.

Solution. The electric field exerts an additional force $(qE \cos \omega t, 0, 0)$ on the particle and the equations of motion (5.10) have therefore to be amended to read:

$$\ddot{x} = \omega \dot{y} + \alpha \cos \omega t, \qquad \ddot{y} = -\omega \dot{x},$$

where $\omega = qB/m$ and $\alpha = qE/m$.

Integrating the second equation under the initial conditions $x = 0$, $\dot{y} = 0$, we get $\dot{y} = -\omega x$ and the first equation then yields $\ddot{x} + \omega^2 x = \alpha \cos \omega t$. This equation has a general solution $x = P \cos \omega t + Q \sin \omega t + (\alpha t \sin \omega t)/2\omega$. At $t = 0$, $x = 0$ and $\dot{x} = 0$; thus $P = Q = 0$. Substituting for x in $\dot{y} = -\omega x$ and integrating under the initial condition $y = 0$ at $t = 0$, y is now found. The complete solution is

$$x = \frac{\alpha t}{2\omega} \sin \omega t, \qquad y = \frac{\alpha t}{2\omega} \cos \omega t - \frac{\alpha}{2\omega^2} \sin \omega t.$$

Putting $\omega t = \theta$, $\alpha/2\omega^2 = \beta$, parametric equations for the trajectory are found in the form $x = \beta\theta \sin \theta$, $y = \beta\theta \cos \theta - \beta \sin \theta$.

A little consideration reveals that these are the equations of a spiral. For large values of θ, the term $\beta \sin \theta$ is small by comparison with $\beta\theta \cos \theta$ and approximate equations for the trajectory are $x = \beta\theta \sin \theta$, $y = \beta\theta \cos \theta$; the corresponding polar equation is $\rho = \beta\phi$ $(\rho = \sqrt{(x^2 + y^2)}$, $\phi = \tan^{-1}(y/x) = \theta)$, which is clearly the equation of a spiral. □

5.6 Maxwell's Equations Thus far in this chapter, it has been assumed that the magnetic field due to a variable current can be calculated by the methods given in Chapter 4 for the case of steady currents. The implication is that equation (4.11) remains valid, even when $\boldsymbol{j}$ varies with t. Such an assumption is clearly unjustified since, if the divergence of both members of this equation is taken, the result is $\text{div}\, \boldsymbol{j} = 0$ ($\text{div curl}\, \boldsymbol{H}$ vanishes, L. Marder, *Vector Fields*, p. 36) and this equation is incompatible with equation (3.4) unless ρ is independent of t (i.e. steady conditions). If, however, the currents vary comparatively slowly, the term $\partial\rho/\partial t$ is negligible by comparison with div $\boldsymbol{j}$, and equation (4.11) can be accepted as an approximation without inconsistency. We say that the conditions are *quasi-steady*.

For general non-steady conditions, Maxwell assumed that the equations

$$\operatorname{div} \boldsymbol{D} = \rho, \qquad \operatorname{div} \boldsymbol{B} = 0, \tag{5.11}$$

remain valid. Then, equation (3.4) can be written

$$\operatorname{div}\left(\boldsymbol{j}+\frac{\partial \boldsymbol{D}}{\partial t}\right) = 0 \tag{5.12}$$

and this is consistent with an amended equation (4.11), namely

$$\operatorname{curl} \boldsymbol{H} = \boldsymbol{j}+\frac{\partial \boldsymbol{D}}{\partial t}. \tag{5.13}$$

$\partial \boldsymbol{D}/\partial t$ is termed the density of the *displacement current*. Maxwell postulated equation (5.13) as a fundamental equation of the electromagnetic field.

The set of fundamental equations was completed by the addition of one representing Faraday's law. If C is any fixed closed curve lying in an electromagnetic field, the e.m.f. in a circuit coincident with C is $\oint_C \boldsymbol{E}.d\boldsymbol{s}$, $d\boldsymbol{s}$ being an element of C. Let S be any surface bounded by C and $\boldsymbol{n}$ the unit normal to S. Then, equation (5.1) can be expressed in the form

$$\int_C \boldsymbol{E}.d\boldsymbol{s} = -\frac{d}{dt}\int_S \boldsymbol{B}.\boldsymbol{n}\, dS = -\int_S \frac{\partial \boldsymbol{B}}{\partial t}.\boldsymbol{n}\, dS.$$

Applying Stokes' theorem to the left-hand integral, we deduce the final Maxwell equation,

$$\operatorname{curl} \boldsymbol{E} = -\frac{\partial \boldsymbol{B}}{\partial t}. \tag{5.14}$$

In a solid material medium, the following equations are also usually valid,

$$\boldsymbol{D} = \varepsilon\varepsilon_0 \boldsymbol{E}, \qquad \boldsymbol{B} = \mu\mu_0 \boldsymbol{H}, \qquad \boldsymbol{j} = \sigma \boldsymbol{E}. \tag{5.15}$$

At a boundary between two media, assuming that the surface densities of charge and current are zero, Maxwell's equations require that the normal components of $\boldsymbol{B}$ and $\boldsymbol{D}$, and the tangential components of $\boldsymbol{H}$ and $\boldsymbol{E}$, must be continuous.

Problem 5.11 An electromagnetic oscillation is taking place inside a perfectly conducting spherical shell of radius a. Taking the centre of the shell as the pole of spherical polar coordinates (r, θ, ϕ), show that Maxwell's equations can be satisfied by taking $E_r = f(r)\cos\theta \exp(j\omega t)$,

$E_\theta = g(r)\sin\theta\exp(j\omega t)$, $E_\phi = 0$, $H_r = H_\theta = 0$, $H_\phi = h(r)\sin\theta\exp(j\omega t)$, provided f, g, h are related by certain differential equations. Putting $f = r^{-\frac{3}{2}}y$, $\omega r/c = x$ ($c^2 = 1/\mu_0\varepsilon_0$), show that y satisfies Bessel's equation and deduce that ω must satisfy the equation $\tan\zeta = \zeta/(1-\zeta^2)$, where $\zeta = \omega a/c$.

Solution. Inside the shell, $\boldsymbol{j}$ and ρ both vanish and $\boldsymbol{B} = \mu_0\boldsymbol{H}$, $\boldsymbol{D} = \varepsilon_0\boldsymbol{E}$. Thus, Maxwell's equations reduce to

$$\operatorname{div}\boldsymbol{E} = \operatorname{div}\boldsymbol{H} = 0, \qquad \operatorname{curl}\boldsymbol{H} = \varepsilon_0\frac{\partial\boldsymbol{E}}{\partial t}, \quad \operatorname{curl}\boldsymbol{E} = -\mu_0\frac{\partial\boldsymbol{H}}{\partial t}. \tag{5.16}$$

Expressing these equations in spherical polar coordinates (L. Marder, *Vector Fields*, p. 64) and substituting the assumed forms for the spherical polar components of $\boldsymbol{H}$ and $\boldsymbol{E}$, it will be found that some equations are satisfied identically and the remainder reduce to the forms

$$\begin{aligned} &\frac{d}{dr}(r^2 f)+2rg = 0,\\ &\frac{2h}{r} = j\omega\varepsilon_0 f,\\ &\frac{1}{r}\frac{d}{dr}(rh) = -j\omega\varepsilon_0 g,\\ &\frac{1}{r}\frac{d}{dr}(rg)+\frac{f}{r} = -j\omega\mu_0 h. \end{aligned} \tag{5.17}$$

Elimination of h between the middle pair of equations yields the first equation; this shows that only three of these equations are independent. Eliminating h from the last equation by use of the second equation, we find that

$$\frac{d}{dr}(rg)+f\left(1-\frac{\omega^2 r^2}{2c^2}\right) = 0. \tag{5.18}$$

Equations (5.17), (5.18) now give

$$\frac{d^2}{dr^2}(r^2 f)-f\left(2-\frac{\omega^2 r^2}{c^2}\right) = 0. \tag{5.19}$$

To solve this equation, put $f = r^{-\frac{3}{2}}y$ and $\omega r/c = x$; this transforms the equation to the form

$$x^2y''+xy'+(x^2-\tfrac{9}{4})y = 0;$$

this is Bessel's equation of order $\frac{3}{2}$ (J. Heading, *Ordinary Differential Equations*, p. 82). Accepting the solution which is bounded at $x = 0$ (i.e. the pole), we get

$$y = J_{\frac{3}{2}}(x) = \left(\frac{2}{\pi}\right)^{\frac{1}{2}} x^{-\frac{3}{2}}(\sin x - x\cos x).$$

The functions, f, g, h now follow immediately:

$$f = r^{-\frac{1}{2}} J_{\frac{3}{2}}\left(\frac{\omega r}{c}\right), \quad g = -\frac{1}{2r}\frac{d}{dr}\left[r^{\frac{1}{2}} J_{\frac{3}{2}}\left(\frac{\omega r}{c}\right)\right], \quad h = \tfrac{1}{2} j\omega\varepsilon_0 r^{-\frac{1}{2}} J_{\frac{3}{2}}\left(\frac{\omega r}{c}\right);$$

an arbitrary multiplier has, of course, been omitted.

Since the shell is infinitely conducting, $\sigma = \infty$ and, thus, $\boldsymbol{j} = \sigma\boldsymbol{E}$ implies that $\boldsymbol{E} = 0$ within the shell. The boundary conditions over $r = a$ are accordingly $E_\theta = E_\phi = 0$, i.e. $g(a) = 0$. This leads to the condition

$$\frac{d}{dx}\left(\cos x - \frac{\sin x}{x}\right) = 0$$

for $x = \zeta$, or $\tan\zeta = \zeta/(1-\zeta^2)$. This last equation determines possible frequencies of oscillation in this mode (the smallest root is $\zeta = 2{\cdot}74$).

The lines of magnetic intensity are circles in planes perpendicular to the axis $\theta = 0$ and the lines of electric force are curves in the meridian planes $\phi =$ constant. Such a device is called a *cavity resonator*. □

Taking the curl of both sides of the last of Maxwell's equations in empty space (equations (5.16)) and employing the third equation, $\boldsymbol{H}$ is eliminated to give $\operatorname{curl}\operatorname{curl}\boldsymbol{E} = -\mu_0\varepsilon_0\,\partial^2\boldsymbol{E}/\partial t^2$. But $\operatorname{curl}\operatorname{curl}\boldsymbol{E} = \operatorname{grad}\operatorname{div}\boldsymbol{E} - \nabla^2\boldsymbol{E}$ (L. Marder, *Vector Fields*, p. 39). Hence, since $\operatorname{div}\boldsymbol{E} = 0$, $\boldsymbol{E}$ satisfies the wave equation

$$\nabla^2\boldsymbol{E} = \frac{1}{c^2}\frac{\partial^2\boldsymbol{E}}{\partial t^2}, \tag{5.20}$$

where $c^2 = 1/\mu_0\varepsilon_0$. c is the velocity of propagation of electromagnetic waves; substitution of the values for ε_0, μ_0 given in §§ 1.1, 2.5, shows that c is very nearly 3×10^8 m s^{-1}. This is the velocity of light in vacuo, indicating that light is an electromagnetic phenomenon. Elimination of $\boldsymbol{E}$, shows that $\boldsymbol{H}$ satisfies the same wave equation. In uncharged, isotropic material, having dielectric constant ε and permeability μ, $\boldsymbol{E}$ and $\boldsymbol{H}$ also satisfy the wave equation, but the propagation velocity is a, where $a^2 = 1/(\varepsilon\varepsilon_0\,\mu\mu_0)$.

A plane, monochromatic (i.e. definite frequency), plane polarized wave being propagated in the direction of the positive z-axis with velocity a in a medium (ε, μ) has components

$$\begin{aligned} \boldsymbol{E} &= \{E_0 \exp[j\omega(t-z/a)], 0, 0\} \\ \boldsymbol{H} &= \{0, H_0 \exp[j\omega(t-z/a)], 0\}, \end{aligned} \tag{5.21}$$

where $(\varepsilon\varepsilon_0)^{\frac{1}{2}}E_0 = (\mu\mu_0)^{\frac{1}{2}}H_0$ (it is understood that E_0, H_0 may be complex and that real parts are to be taken). The reader should verify that Maxwell's equations are satisfied. A similar wave being propagated in the opposite sense is determined by

$$\begin{aligned} \boldsymbol{E} &= \{E_0 \exp[j\omega(t+z/a)], 0, 0\} \\ \boldsymbol{H} &= \{0, -H_0 \exp[j\omega(t+z/a)], 0\}, \end{aligned} \tag{5.22}$$

where $(\varepsilon\varepsilon_0)^{\frac{1}{2}}E_0 = (\mu\mu_0)^{\frac{1}{2}}H_0$.

Problem 5.12 The region $z > 0$ is occupied by a uniform conductor of conductivity σ and dielectric constant ε and the region $z < 0$ is empty. The wave determined by equations (5.21) is incident normally upon the face $z = 0$ of the conductor. Assuming this gives rise to a field in the conductor determined by

$$E_x = E\exp(j\omega t - kz), \qquad E_y = E_z = 0,$$
$$H_x = 0, \qquad H_y = H\exp(j\omega t - kz), \qquad H_z = 0,$$

and a reflected wave in the vacuum of the form (5.22) and assuming $\sigma \gg \omega\varepsilon\varepsilon_0$, show that the ratio of the amplitudes of the reflected and incident waves is

$$\left[\frac{1+\lambda-\sqrt{(2\lambda)}}{1+\lambda+\sqrt{(2\lambda)}}\right]^{\frac{1}{2}},$$

where $\lambda = \omega\mu\varepsilon_0/\sigma$.

Solution. Setting $\rho = 0$, $\boldsymbol{j} = \sigma\boldsymbol{E}$, $\boldsymbol{D} = \varepsilon\varepsilon_0\boldsymbol{E}$, $\boldsymbol{B} = \mu\mu_0\boldsymbol{H}$ in Maxwell's equations, these become

$$\begin{aligned} \operatorname{div}\boldsymbol{E} &= \operatorname{div}\boldsymbol{H} = 0, \\ \operatorname{curl}\boldsymbol{H} &= \sigma\boldsymbol{E} + \varepsilon\varepsilon_0\,\partial\boldsymbol{E}/\partial t, \\ \operatorname{curl}\boldsymbol{E} &= -\,\mu\mu_0\,\partial\boldsymbol{H}/\partial t. \end{aligned}$$

These are valid in the conductor. It is found that these equations are satisfied by the assumed forms of $\boldsymbol{E}$ and $\boldsymbol{H}$ provided E and H satisfy

$$(\sigma + j\omega\varepsilon\varepsilon_0)E - kH = kE - j\omega\mu\mu_0 H = 0.$$

For these equations to have non-zero solutions in E and H, it is necessary that

$$\begin{vmatrix} \sigma + j\omega\varepsilon\varepsilon_0 & -k \\ k & -j\omega\mu\mu_0 \end{vmatrix} = 0,$$

i.e. $k^2 = \omega\mu\mu_0(j\sigma - \omega\varepsilon\varepsilon_0)$. If $\sigma \gg \omega\varepsilon\varepsilon_0$, this gives $k = +(\frac{1}{2}\omega\mu\mu_0\sigma)^{\frac{1}{2}}(1+j)$ (N.B. positive root must be taken so that the field vanishes as $z \to +\infty$.) With this value of k, the ratio $H:E$ is determined by the equation $H/E = [\sigma/2\omega\mu\mu_0)]^{\frac{1}{2}}(1-j)$.

Taking the reflected wave to be given by

$$\boldsymbol{E} = [E'_0 \exp\{j\omega(t+z/c)\}, 0, 0],$$
$$\boldsymbol{H} = [0, -H'_0 \exp\{j\omega(t+z/c)\}, 0],$$

the net disturbance in the region $z < 0$ is found by superposing this wave on the incident wave (5.21) (with $a = c$). Thus, the field in the vacuum at $z = 0$ is given by

$$\boldsymbol{E} = [(E_0 + E'_0)\exp(j\omega t), 0, 0], \qquad \boldsymbol{H} = [0, (H_0 - H'_0)\exp(j\omega t), 0].$$

The field in the conductor at $z = 0$ is

$$\boldsymbol{E} = [E\exp(j\omega t), 0, 0], \qquad \boldsymbol{H} = [0, H\exp(j\omega t), 0].$$

Since E_x, H_y must be continuous across $z = 0$, the boundary conditions are

$$E_0 + E'_0 = E, \qquad H_0 - H'_0 = H.$$

The second of these conditions implies that

$$\left(\frac{\varepsilon_0}{\mu_0}\right)^{\frac{1}{2}}(E_0 - E'_0) = \left(\frac{\sigma}{2\omega\mu\mu_0}\right)^{\frac{1}{2}}(1-j)E.$$

We can now eliminate E and solve for the ratio $E'_0 : E_0$; the result is

$$\frac{E'_0}{E_0} = \frac{(2\varepsilon_0\omega\mu)^{\frac{1}{2}} - \sigma^{\frac{1}{2}}(1-j)}{(2\varepsilon_0\omega\mu)^{\frac{1}{2}} + \sigma^{\frac{1}{2}}(1-j)}.$$

The amplitudes of the incident and reflected waves are $|E_0|$ and $|E'_0|$ respectively; hence, taking moduli, the result stated now follows.

In non-ferromagnetic materials, $\mu = 1$ and λ will therefore be small. The ratio then approximates to $1 - \surd(2\lambda)$. □

EXERCISES

1. A current i is maintained in the long straight wire shown in Fig. 4.6 If the centre of the circular wire is made to oscillate in the line OA with simple harmonic motion of small amplitude ε and angular frequency ω, its mean distance from the straight wire being c, show that an alternating electromotive force of amplitude $\mu_0 i\varepsilon\omega[1 - c/\surd(c^2 - a^2)]$ is induced within it.

2. Show that the mutual inductance between a square circuit of side $2a$ and a long straight current lying in the plane of the square parallel to one

of its sides and distant c from its centre ($c > a$), is

$$\frac{\mu_0 a}{\pi}\log\left(\frac{c+a}{c-a}\right).$$

3. The primary and secondary circuits of a transformer are identical and have resistance R and self-inductance L. Their mutual inductance is L/k ($k > 1$). At $t = 0$, the primary terminals are connected to a source of constant e.m.f. E and the secondary terminals are short circuited. Show that the maximum current which subsequently flows in the secondary is

$$\frac{E}{R}\cdot\frac{(k-1)^{\frac{1}{2}(k-1)}}{(k+1)^{\frac{1}{2}(k+1)}}.$$

If the coupling is tight, so that k approximates to 1, prove that the peak current is almost $E/2R$.

4. The primary and secondary windings of a transformer have self-inductances L, L' respectively and negligible resistance. A resistance R and capacitance C are connected in series across the secondary terminals and an e.m.f. of amplitude E and angular frequency ω is applied to the primary terminals. If M is the mutual inductance between the windings, show that the current in the secondary is a maximum when $\omega^2 = L/C(LL'-M^2)$ and that the equivalent impedance across the supply is then LR'/M.

5. The space between two very long, circular, coaxial cylinders is evacuated and electrons are set free with negligible initial velocity at the surface of the inner cylinder. They are accelerated from the inner cylinder, of radius a, towards the outer, of radius b, by means of a constant potential difference V between the cylinders. If a uniform magnetic field B is applied parallel to the axis of the cylinders, show that the electrons will just reach the outer cylinder if

$$8mb^2V = eB^2(b^2-a^2)^2,$$

e being the charge and m the mass of the electron. (*Magnetron*)

6. Repeat the cavity resonator calculation of Problem 5.11 for an oscillatory mode in which

$$H_r = f(r)\cos\theta\exp(j\omega t), \qquad H_\theta = g(r)\sin\theta\exp(j\omega t), \qquad H_\phi = 0,$$

$$E_r = E_\theta = 0, \qquad E_\phi = h(r)\sin\theta\exp(j\omega t)$$

and show that the possible frequencies ω are determined by the equation $\tan\zeta = \zeta$.

Appendix

International System of Units (Rationalized MKSA System)

Quantity	*SI unit*
Length	metre (m)
Mass	kilogramme (kg)
Time	second (s)
Force	newton (N)
Work and energy	joule (J)
Power	watt (W)
Charge	coulomb (C)
Current	ampère (A)
Potential	volt (V)
Electric dipole moment	coulomb-metre
Electric displacement $\boldsymbol{D}$	coulomb-metre^{-2}
Electric intensity $\boldsymbol{E}$	newton-coulomb^{-1} or volt-metre^{-1}
Electric polarization $\boldsymbol{P}$	coulomb-metre^{-2}
Flux of magnetic induction N	weber (Wb)
Magnetic induction $\boldsymbol{B}$	weber-metre^{-2} or tesla (T)
Magnetic intensity $\boldsymbol{H}$	ampère-metre^{-1}
Magnetic pole strength p	ampère-metre
Magnetic moment $\boldsymbol{m}$	ampère-metre^2
Magnetization $\boldsymbol{M}$	ampère-metre^{-1}
Capacitance C	farad (F)
Resistance R	ohm (Ω)
Inductance L	henry (H)
Frequency $\nu = \omega/2\pi$	hertz (Hz)

Index